NOMADIC AND INDIGENOUS SPACES

Nomadic and Indigenous Spaces

Productions and Cognitions

Edited by

JUDITH MIGGELBRINK
Leibniz Institute for Regional Geography, Germany

JOACHIM OTTO HABECK
Max Planck Institute for Social Anthropology, Germany

NUCCIO MAZZULLO
Leibniz Institute for Regional Geography, Germany and Arctic Centre, University of Lapland, Finland

PETER KOCH
Leibniz Institute for Regional Geography, Germany

Routledge
Taylor & Francis Group
LONDON AND NEW YORK

First published 2013 by Ashgate Publishing

2 Park Square, Milton Park, Abingdon, Oxon OX14 4RN
711 Third Avenue, New York, NY 10017, USA

Routledge is an imprint of the Taylor & Francis Group, an informa business

First issued in paperback 2016

The views of individual authors do not necessarily reflect those of the editors.

British Library Cataloguing in Publication Data
A catalogue record for this book is available from the British Library

The Library of Congress has cataloged the printed edition as follows:
Nomadic and indigenous spaces : productions and cognitions / by Judith Miggelbrink, Joachim Otto Habeck, Nuccio Mazzullo and Peter Koch, [editors]
p. cm.
Includes bibliographical references and index.
ISBN 978 1 4094 6458 7 (hardback)
1. Nomads. 2. Indigenous peoples. 3. Human geography. 4. Spatial behavior--Computer network resources. 5. Geographical perception--Computer network resources. 6. Mobile geographic information systems. I. Miggelbrink, Judith.
GN387.N592 2013
305.9'06918--dc23

2013009607

ISBN 978-1-4094-6458-7 (hbk)
ISBN 978-1-138-26721-3 (pbk)

Contents

List of Figures

List of Maps

Notes on Contributors

Claudio Aporta was born and raised in the province of Mendoza, Argentina. He received his Ph.D. from the University of Alberta, Canada. His current affiliation is Associate Professor, Marine Affairs Program, Dalhousie University, Halifax, Canada. His major research interests are connected to how Inuit relate to their physical environment as well as the transmission of Inuit oral knowledge in contemporary contexts.

Brian Donahoe (Ph.D. Indiana University, 2004) is an independent researcher, writer, and editor living in Kyzyl, Republic of Tyva, Russia. From 2004 to 2010 he was post-doctoral research fellow at the Max Planck Institute for Social Anthropology in Halle, Germany. His research interests include the dynamics of constructing, maintaining, and asserting ethnic identity and indigeneity, and different approaches to guaranteeing indigenous rights to land. He is co-editor of two recent volumes: *Law Against the State: Ethnographic Forays into Law's Transformations* (Cambridge, 2012) and *Reconstructing the House of Culture: Community, Self and the Makings of Culture in Russia and Beyond* (Berghahn, 2011).

Gail Fondahl is Professor of Geography at the University of Northern British Columbia, Canada. She holds a Ph.D. in geography from the University of California Berkeley, USA. Her interests lie in indigenous rights, especially to land and resources in the Russian and Canadian Norths, resource co-management, research co-management, and indicators of human development in the North.

Joachim Otto Habeck is Coordinator of the Siberian Studies Centre at the Max Planck Institute for Social Anthropology in Halle, Germany. He received his Ph.D. from the University of Cambridge in 2004. He is author of *What it Means to Be a Herdsman: The Practice and Image of Reindeer Husbandry among the Komi of Northern Russia* (2005). He has conducted research on mobile pastoralism in the Russian North, the history of sedentarization in Central Siberia, and the involvement of cultural institutions in reshaping ethnic identities. With Brian Donahoe, he co-edited *Reconstructing the House of Culture* (Berghahn, 2011).

Tim Ingold is Professor of Social Anthropology at the University of Aberdeen, Scotland, and a Fellow of both the British Academy and the Royal Society of Edinburgh. He received his Ph.D. in Social Anthropology in 1976 from the University of Cambridge. He has carried out ethnographic fieldwork among Skolt Sámi and Finnish reindeer herders in Finnish Lapland and conducted research on environment, technology, human-animal relations, and social organization in the circumpolar North including northern America and Siberia. Tim Ingold made his mark especially in comparative and evolutionary anthropology, and subsequently in the study of skilled practice and environmental perception. He was the editor of *Man* (1990–92) and of the *Companion Encyclopedia of Anthropology* (1994). His books include *Evolution and Social Life* (1986), *The Appropriation of Nature* (1986), *Tools, Language and Cognition in Human Evolution* (co-edited with Kathleen Gibson, 1993), *Key Debates in Anthropology* (1996), *The Perception of the Environment* (2000), *Lines: A Brief History* (2007), *Creativity and Cultural Improvisation* (co-edited with Elizabeth Hallam, 2007), *Ways of Walking* (co-edited with Jo Lee Vergunst, 2008), *Being Alive* (2011), *Redrawing Anthropology* (2011), *Imagining Landscapes* (co-edited with Monica Janowski 2012), *Biosocial Becomings* (co-edited with Gísli Pálsson, 2013), and *Making* (2013). Ingold is currently writing on issues on the interface of anthropology, archaeology, art, and architecture.

Kirill V. Istomin grew up in Syktyvkar, currently the Komi Republic, Russian Federation. After graduating from the University of Tartu, Estonia in 2001, he worked for the Komi Science Centre, Ural Division, Russian Academy of Sciences, in Syktyvkar. In 2005 he earned his Candidate of Sciences degree (equivalent to a Ph.D.) from the Institute of Ethnology and Anthropology, Russian Academy of Sciences, in Moscow. He has been working at the Max Planck Institute for Social Anthropology in Halle, Germany, from 2005 to 2012. Having returned to Syktyvkar, he is now senior researcher at the Komi Science Centre. His academic interests include spatial orientation and other cognitive skills of nomads, ecology and technology of reindeer herding, cultural psychology, and neuroanthropology.

Peter Koch is a Ph.D. student at the Leibniz Institute for Regional Geography in Leipzig, Germany. He has a regional focus on Scandinavia, especially Sápmi, and the Baltic Sea Region and works on society and spatiality with a major interest in cross-border reindeer herding and national park policies.

Joseph J. Long is Research Associate in the Siberian Studies Centre at the Max Planck Institute for Social Anthropology in Halle, Germany, and currently Researcher in Residence with Scottish Autism. He has undertaken long-term fieldwork in Buryat communities in the Baikal region of Siberia where his research interests include ritual and performance, kinship, civic cultural institutions, and the politics of indigeneity. He received his Ph.D. from the University of Aberdeen in 2010.

Nuccio Mazzullo has a Ph.D. in social anthropology from the University of Manchester, England. Currently he is Researcher at the Leibniz Institute for Regional Geography in Leipzig, Germany, and Fellow Researcher at the Arctic Centre, University of Lapland, Rovaniemi, Finland. His regional specialization is Finnish Lapland. Topical interests include reindeer herding, human–environment relationships, environmental politics and indigenous rights, landscape and perception, movement and place naming, learning skills and handicraft, narratives and identity, and the anthropology of circumpolar peoples.

Judith Miggelbrink has a Ph.D. in geography from the University of Leipzig and is head of a research group at the Leibniz Institute for Regional Geography in Leipzig, Germany. She has led and carried out several research projects concerning the role of spatiality and territoriality in society, especially on the effects of borders in everyday life and practices. Her regional focus is on Northern, Central and Eastern Europe.

Denis Retaillé is Professor of Geography at the University of Bordeaux III, France, and the editor of the French journal *L'Information géographique*. He is the former president of the French Association for the Development of Geography. After early works on nomadic and sedentary peoples' encounters in Sahel and South Sahara, he created a model of the globalized world using the nomadic conceptualization of space, which he introduced in his *Le monde, espaces et systèmes* (with Lévy and Durand). In *Le monde du géographe* he defined novel basic concepts – especially *place* – in order to understand the world from a mobile point of view. Following his *Concepts du nomadisme, nomadisation des concepts* Denis Retaillé started to advocate a *mobile space theory*. It both rejects a metaphoric use of *nomadic space* and represents a new contractual space, which with regard to the end of social contract – the fictional basis of the modern state – needs further investigation.

Florian Stammler is Coordinator of the Anthropology Research Team at the Arctic Centre, University of Lapland, Rovaniemi, Finland, and also an institute associate at the Scott Polar Research Institute of the University of Cambridge. He has led research projects and published extensively on reindeer herding, extractive industry and oral history of northern residents. In the period 2010–15 he leads a comparative research project on Oral Histories of Empires by Elders in the Arctic (ORHELIA).

Denis Wood is an independent scholar who holds a Ph.D. in geography from Clark University, Worcester, MA, United States. He is the author of ten books and 150 papers. From 1974 to 1996 he taught in the School of Design at North Carolina State University. He curated the exhibition *The Power of Maps* for the Cooper Hewitt National Museum of Design in 1992 and remounted it at the Smithsonian in Washington in 1994. Accompanying the exhibition was Wood's extremely popular and widely cited book of the same name. In 2008 the University of Chicago Press came out with his *The Natures of Maps*, co-authored with John Fels; and in 2010 Guilford Press published his *Rethinking the Power of Maps*, as well as a second edition of *Making Maps*, co-authored with John Krygier. Since then Siglio Press has published his neighbourhood atlas, *Everything Sings*, maps from which have been exhibited internationally.

Acknowledgements

The contributions to this book have emerged from a conference with the title 'Nomadic and Indigenous Spaces: Productions and Cognitions' conducted by Judith Miggelbrink (Leibniz Institute for Regional Geography, Leipzig) and Günther Schlee (Max Planck Institute for Social Anthropology, Halle) at the University of Leipzig in February 2011.

The conference was organized with the support of the Collaborative Research Centre *Difference and Integration – Interaction between Nomadic and Settled Forms of Life in the Civilizations of the Old World*. This Centre, also known by its short label SFB 586, brought together scholars from many disciplines to elucidate forms, impacts and interpretations of interactions between nomadic and sedentary people in history and present. The Collaborative Research Centre was funded by the German Research Foundation (Deutsche Forschungsgemeinschaft) from 2001 to 2012 (see www.nomadsed.de). In particular, the conference was based on a collaboration of two projects both exploring the role of space in nomadic and indigenous ways of living: *New Technologies in the Tundra: High-Tech Equipment, Perception of Space and Spatial Orientation of Nomadic and Settled Populations of the Russian Arctic* conducted by Günther Schlee, Kirill Istomin and Joachim Otto Habeck at the Max Planck Institute for Social Anthropology, and *Power Technologies' Productions of Space: Sámi Territoriality and Indigeneity* conducted by Judith Miggelbrink, Nuccio Mazzullo and Peter Koch at the Leibniz Institute for Regional Geography.

We would like to express our special thanks to the German Research Foundation (Deutsche Forschungsgemeinschaft) and the Collaborative Research Centre SFB 586. We highly appreciate the inspiration and constructive criticism that we received from our colleagues at the Centre. The volume would not have been possible without the active support of many other people: We are grateful to Bernd Belina from the University of Frankfurt (Main) for accompanying the project on power technologies and especially for his cooperation in setting up the conference. We thank Valerie Rose, Carolyn Court, and Philip Stirups at Ashgate for managing the publishing process and for patiently answering all our queries.

We would also like to express our gratitude to two anonymous reviewers for their effort to scientifically validate the concept of the volume. We are grateful to Silke Dutzmann from the Leibniz Institute for Regional Geography, who designed all maps and figures and from whose cartographic experience and professional knowledge we all benefitted; and to two student assistants, Angelika Brünecke and Tom Schwarzenberg, who supported the editing of the manuscripts from the initial to the final stage of the volume. We would also like to express our gratitude to Deborah Connolly from LSI World of Languages, Leipzig, for her persevering copy-editing. Finally, our most sincere thanks go to the authors and their interlocutors, whose views, insights and stories provide the basis for this intellectual expedition. You have made this journey possible.

Chapter 1

Nomadic and Indigenous Spaces: Paths and Perspectives

Judith Miggelbrink, Joachim Otto Habeck, Nuccio Mazzullo and Peter Koch

This is the age of smartphones and internet-based maps. If you want to get from one place to another, digital technology makes it easy to find your way. There is no need to pester locals with questions, no need to unfold a printed map or thumb through a timetable book. Directions are clear. We all know how to use these devices. 'Turn left after 200 metres. You have reached your destination'. All this works perfectly – provided there is electricity – in most parts of the United States and Europe, at least. Yet in many other parts of the world, relying on iPhone and Google Maps would be foolish. It does not really make sense to ask your iPhone for directions to Boultoum in Niger (see Chapter 3) or Olenek in Siberia (Figure 1.1, see Chapter 8). For what the smartphone instructions from Leipzig to Olenek do not mention is that the last few hundred kilometres are winter roads that run across ridges and over frozen rivers. Even in a suitable vehicle, it is impossible to reach the place without local knowledge. You would not travel on your own to these places, but rely on collective means of transportation, personal contacts established in advance and the assistance of local travellers. Where snow, frozen rivers, or sand are the surfaces that facilitate movement, the predictability of transportation known from Western settings, measured in hours and minutes of estimated travel time, no longer holds. Instead, moving involves waiting: waiting for the right time, the right moment to depart, the right weather, the right person to travel with. It is about taking lifts and taking chances.

The title of this volume, *Nomadic and Indigenous Spaces*, is shorthand for our central proposition that spatiality has its distinct properties when comprehended in nomadic or indigenous contexts. Rather than demarcating nomadic from sedentary spaces or indigenous spaces from settlers' spaces, the idea is to highlight and analyse the ways members of nomadic and indigenous communities perceive space, navigate in space, move and dwell, entertain relations with that what surrounds them, leave their own marks in the landscape, carry the marks of living in a certain place, take that place with them, tell their own stories and produce their own meanings of space. *Indigenous* and *nomadic* will be juxtaposed below; their meaning may be legally fixed or simply hinge on *ad-hoc* identifications, and these two words play out their significance mostly in encounters with a hegemonic or sedentary Other.

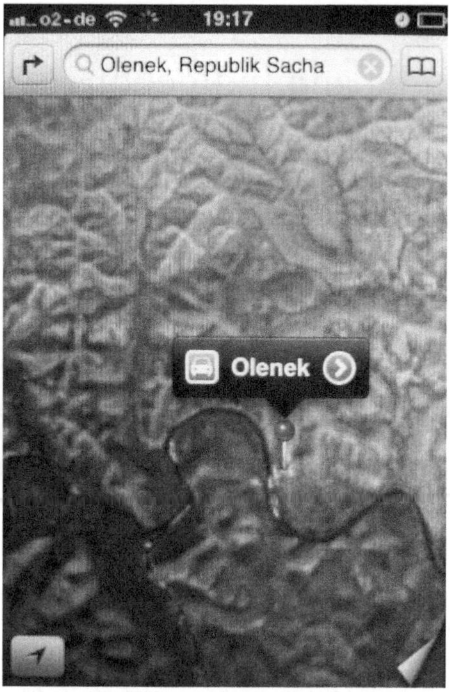

Figure 1.1 Leipzig to Olenek: 9,390 kilometres, 6 days and 12 hours by car, according to iPhone maps and instructions

Source: Joachim Otto Habeck, 1 December 2012

It is not our intention to romanticize the *indigenous* and *nomadic*. Returning to the above example of movement, newly invented technologies, such as GPS or mobile phones, have without any doubt a strong bearing on livelihoods and everyday practices in all modes of life, be they nomadic or sedentary: they create and sustain new forms of togetherness as much as personal accountability, they enhance multi-level communication as much as top-down surveillance. In addition, and often in tandem with technological innovations, there are shifting or newly emerging scales of social action and interaction, such as transnational arenas. Not only do these changes affect people's spatial organization and orientation, they also challenge subjective and collective identifications.

The editors of this volume come from different backgrounds, they have their scientific roots in different disciplines – social anthropology and geography – and followed different interests when starting to investigate nomadic and indigenous spaces. They have worked in two scientific projects of one research programme.[1]

1 See the Notes on Contributors at the beginning of this volume. The project *New Technologies in the Tundra: High-Tech Equipment, Perception of Space and Spatial Orientation of Nomadic and Settled Populations of the Russian Arctic* (2008–12, Schlee, Istomin, Habeck)

They decided to join forces to combine two sets of approaches to space that usually stand apart: the volume's subtitle renders these as *production* and *cognition*. On the one hand, there are production of space approaches that help explain how, by whom and to what ends space is produced within certain social conditions and how it might have effects on them. Cognition of space approaches, on the other hand, focus on processes of perception and navigation, their dynamics and stabilities. Both perspectives will be introduced in more detail below. The aim of this volume is to employ both perspectives to assess how nomadic and indigenous groups go about space and make a living in changing political and technological settings.

Most contributors present case studies from the North – notably, Canada, northern Europe, and Siberia – which is owed to the regional research foci of the editors. These case studies are complemented by one West African example (Retaillé's chapter) and rounded off by three general contributions (those of Donahoe, Wood and Ingold). The individual chapters are summarized at the end of this introduction. First, however, we draw comparisons on how the concepts of space, spatial cognition, and production of space have been treated in the two disciplines that we represent – anthropology and geography. We continue with a brief juxtaposition of *nomadic* and *indigenous*, take a look at Deleuze's and Guattari's (1987) idea of 'smooth space' (or 'nomad space') versus 'striated space', and then sketch out some strands we consider to be particularly productive for transdisciplinary research on spatiality in nomadic and indigenous settings.

Space for Anthropologists, Space for Geographers

Since our authors are geographers and social or cultural anthropologists, we briefly summarize how in these two disciplines the meaning of space has changed over time. One hundred years ago it was valid to claim that geography is the study of the Earth – more precisely, its surface – and anthropology the study of the peoples thereon. Both disciplines involved long-distance travels and expeditions.[2] Geography had the task of making inventories and describing all sorts of objects in their spatial distribution or spatial dimension. Mapping had a heuristic function in generating questions concerning location and distribution, but was also the prime technique of analysis and visualization of research findings. Anthropologists were to examine and map cultural diversity. Along with other factors (political

and the project *Power Technologies' Productions of Space: Sámi Territoriality and Indigeneity* (2008–12, Miggelbrink, Mazzullo, Koch) were both conducted within the *Collaborative Research Centre Difference and Integration: Interactions between Nomadic and Settled Forms of Life in the Civilizations of the Old World* (SFB 586) financed by the German Research Foundation. The contributions to this volume were presented at a conference in Leipzig in February 2011, organized by Judith Miggelbrink and Günther Schlee.

2 Though in geography, travel and expeditions only became popular in the nineteenth century. We are grateful to Dirk Hänsgen for his valuable comments on this point.

conditions in the age of colonialism and the powerful scientific paradigm of evolutionism), it was the mere distance in space that made anthropologists perceive and construct their objects of research as the Other. Space in anthropology was what isolated the researchers from their objects. *Naturvölker* were attributed their own spaces, different and distant from that of *Kulturvölker*. By the same token, ethnographies, as descriptions of indigenous and nomadic peoples, usually started with a description of the space that the examined group populate, and discussed at length the ways in which that group made use of their natural environment.

In *anthropological* scholarship, space has thus been credited with high significance, first and foremost as a stage, a spatial setting in which this or that people earns its livelihood. Yet in addition to portraits of land and resource use, what we also find in many ethnographies are data on labour division and the social partitioning of space (both inside buildings and outside) and the symbolic order of space, on different scales from the interior of the dwelling to entire cosmologies (for Siberia: Anisimov 1963, Harva 1938). It is in these data that indigenous productions of space become elicit. Ethnographic publications themselves, however, also constitute distinct narratives of indigenous spaces insofar as descriptions and representations of indigenous life-worlds are always necessarily selective and arbitrary. Moreover, by locating indigenous peoples in distant spaces, ethnographic accounts also contributed to the discursive construction of a hierarchically conceived global space.

Ethnographies often point to conflicts between different groups, indigenous and non-indigenous, over land and natural resources, and thus conflict over space. In the wake of colonialism and with the formation of indigenous movements, anthropologists increasingly directed their attention to conflicts between the state and native populations. To give a few examples from the North, anthropologists and colleagues from related disciplines became confronted with debates around the construction of the Trans-Alaska Pipeline System in the 1960s and 1970s (Berger 1985), the James Bay Project in Quebec in the 1970s and 1980s (Feit 2000), the Alta hydroelectric power plant in northern Norway around 1980 (Paine 1996), and the Baikal-Amur Mainline Railway at about the same time (Anderson 1991, see also Povorozniuk 2011). More recently, a focus has been placed on conflicts over indigenous rights to marine resources (for example, Scott 2004, Scott and Mulrennan 2010). Conflicts over land and water (space) and natural resources occur throughout the world, but they acquire a particularly problematic character in nomadic settings, where entitlements to land and resources often contradict sedentary bureaucrats' ideas of discrete, neatly bounded and fixed territories and of property in *real estate*.

As a component of the descriptions of different peoples' natural environment and land use, anthropologists also frequently investigated their informants' orientation in space. For example, in the late 1960s and the 1970s there was a remarkable interest in wayfinding strategies of marine navigators in the Pacific (see the section on Spatial Cognition below). Practices of map-making and map-reading also became a subject of anthropological research (Gell 1985). However, it was only in the 1990s that anthropologists turned to explicitly problematizing concepts of space and place, and at about the same time some scholars embarked on constructivist interpretations of space, as will be laid out below (in the section on Spatial Production).

In geography, space has not only been credited with a high significance but has always been treated as a core concept of geographic approaches to the natural as well as to the social world (for example, Livingstone 2002: 7–42, Mayhew 2011, Withers 2011). This strand of reasoning (*Staatenkunde, Länderkunde*) was closely related to the formation of dominions and nation states, thus focusing on its natural endowments and human development in terms of spatial orders of population, agriculture, manufacturing, settlements, infrastructure etc.[3] Equally influential was an entirely different understanding of space – space as landscape – resulting from the human adaption to and appropriation of nature. This has been continued and broadened by geographers who committed themselves to the study of environmental challenges and human responses, conceived as components of human-environment ecological systems (Haggett 1972). However, a growing prevalence for systems, organization, relations, diffusion etc. solidified an abstract and technical understanding of space. This came along with and was based on new *technical possibilities* that could be applied to mapping and map generation, and on *methodological preferences* for modelling, mathematical calculus and statistical tools (for example, Chorley and Haggett 1967) as well as on correspondent *theoretical underpinnings* provided mainly by critical rationalism (Harvey 1969). In contrast to anthropology with its aim of understanding different and distant cultures and ways of living, geography has increasingly been driven by a self-understanding of an applied science, the task of which is to serve planning purposes (notably, physical and regional planning). This, in turn, has strengthened spatial concepts deriving from the political-administrative-hierarchical organization of the state and its territory. Parallel with anthropology, geography has thus contributed to a hegemonic construction of a patterned and hierarchized global space. Moreover, it also contributed to a reified understanding of space as a dimension that does not derive from human activity but precedes it as a putatively neutral stage or container.

Alternative approaches were developed by radical and Marxist geographers (Peet 1985, 1998: 76–111) who have emphasized the socially produced and thus instrumental nature of space for a long time. This will be addressed in more detail below. Beyond this, fixed patterns of geographical reasoning were challenged when geographers became involved in the *cultural turn* (Barnett 1998, Castree 1999, Cook et al. 2000).[4] This has raised new debates about the understanding of culture in geography (for example, Cosgrove 1983, Price and Lewis 1993, Sayer 2000) by which totalizing and 'superorganic' concepts of culture have been called

3 See, for example, the history of geography in Germany (Schultz 2008: 70–76, Schelhaas and Hönsch 2001).

4 Which is – outside geography – also regarded as a spatial turn.

into question, rejected and largely abandoned (Mitchell 1995). Here an interesting symmetry occurs to what has been said about the field of anthropology before: Whereas anthropology has been facing a new reflexivity on space, geography has been confronted with knowledge-based concepts of culture emphasizing the role of symbols, representation, semiotic processes and performativity (Agnew 1997, Nash 2000, Smith and Katz 1993). Moreover, the cultural turn has also initiated a debate on new understandings of space and spatiality. Although strands differ remarkably in detail, they share an understanding of space that reflects its socially constructed and instrumental nature. This comes along with a growing interest in the subject and subjectivity, on the one hand, and discourse, on the other (Gibson 2001, Pile 2008).

Within geography, a substantial body of research on the co-evolution of objective and subjective, material and discursive, and abstract and concrete spatialities sprouted from a critical reading of social theory (for example, Crang and Thrift 2000, Gregory 1998, Harvey 2000a [1989]). Moreover, very recently and after a period of disregard, a new debate on territory has emerged (for example, Antonsich 2010, Elden 2007a–c, 2010), fathoming territoriality as a technology of power. Taking the ontological and epistemological dimensions of social-reality research (Woodward, Jones and Marston 2010) into due account, there is now a reinvigorated discussion on the role of the state as well as its spatial and territorial means in framing everyday practices, and the ways that everyday practices in turn reframe those means. While the theoretical outcomes of these debates are remarkable, studies applying the concepts to empirical investigations are underdeveloped. Our volume is an attempt to close the lacuna between advanced theoretical discussions, on the one hand, and comprehensive empirical investigations, on the other; as all authors illuminate their theoretical abstractions by rich fieldwork.

Spatial Cognition in Anthropology and Geography

In anthropology, interest in the cognition of space emerged from various roots. As mentioned, non-Western peoples' orientation in space has periodically attracted attention in anthropological studies. Istomin and Dwyer (2009: 29) list the pertaining publications.[5] Istomin (personal communication) also points to the shortcomings in this line of inquiry: having branched off from an interdisciplinary endeavour

5 Istomin's and Dwyer's (2009: 29) list reads thus: 'Many studies have focused on peoples who were thought to have extraordinary or even "superhuman" orientation skills: Micronesian mariners (Gladwin 1964, 1970; Lewis 1972), south African Bushmen-San (Porteus 1937; Reuning 1988; Widlok 1997), Australian aboriginals (Lewis 1976a, 1976b), and North American Inuit (Nelson 1969; Carpenter 1973; Aporta 2004; Aporta and Higgs 2005), to name just a few'. We thank Kirill Istomin for his valuable comments on a draft of this introduction.

in the mid-twentieth century, the further pursuit of this topic in anthropology developed in almost complete isolation from similar research in psychology and geography. Nonetheless, anthropology also has the potential to close certain gaps in the study of spatial cognition as pursued by other disciplines: while the latter find it difficult to test for cultural background and personal experience of participants in quantitative studies, these very aspects can be usefully researched with the help of well-established anthropological methods (namely through detailed studies in *the field* rather than in the artificial setting of the laboratory).

Exactly this rapprochement is currently underway. In combination with theories about mental maps borrowed from geography (see below) and cognitive studies (Hutchins 1995), anthropological research on spatial cognition has taken a new direction. It examines navigational strategies among nomadic populations and the role of new technologies in their wayfinding (see below). A related strand of research has been pursued by scholars at the interstice of anthropology and psycholinguistics (Levinson 2003, Widlok 1997). While it has been argued by some anthropologists – notably Ingold (2000) – that human perception of the environment is not conditional on high-level cognitive processes such as representation and decision-making, others have sought to draw exactly on cognitive sciences and neurology to understand how people employ and rely on their different senses in the process of moving or travelling. Even those who treat navigational strategies as 'hard-wired' admit that individual preferences for and proficiency in any of them are subject to learning and practical experience: navigational skills are not just an outcome of certain features in the internal structure of the individual's brain, but they also appear to shape the brain (see Donahoe, this volume).

Geography's interest in questions of perception and cognition dates back to its engagement in behavioural studies; it had its climax in the 1970s when a couple of studies on spatial behaviour (for example, Cox 1972, Cox and Golledge 1981, Gold 1980) as well as on mental maps, images, cognitive maps, spatial schemes, and related topics (for example, Downs and Stea 1973, Gould 1972, Gould and White 1974, Lynch 1960) were published. Perception (and the way cognition and perception are related) has gained influence on geographical reasoning since then (for example, Blaut et al. 2003, Golledge 1999, 2000, Kitchin and Blades 2002). Although studies have varied remarkably in terms of theoretical postulates and methodological choices, it can be stated that by and large they have followed the assumption that *behaviour* – including spatial behaviour – is influenced and, moreover, governed by *perception* – including spatial perception. The underlying premise is, as Wießner (1978: 423) put it, that 'behavior in space' could be explained by a person's individual spatial scheme. This fundamental though general hypothesis has required clarification, especially concerning the varying, sometimes even contradictory references to *space* and the *spatialities* of behaviour and perception.

In geography, perception and cognition have been closely associated with maps and mapping. Mental maps as well as mental mapping, thus, are assumed to play a crucial role in connecting (f)actual doings, perception and cognition. However, mental maps (and mental mapping) have played an iridescently methodological role: In a first and still widely echoed understanding of the term, mental maps apply to the outcome of a methodological process measuring individual or group-related knowledge or preferences about the environment that are then projected onto standardized, earth-grid related geographical maps. A mental map is thus a map constructed according to scientific standards of transparency and comprehensibility (Gould and White 1974: 51–2). Besides a methodologically controlled representation of place-related knowledge *converted into* a map, mental maps also apply to maps or map-like sketches produced in everyday life for a variety of purposes. They represent the logics, purposes and requirements of people, thus being a 'representation of a logical space, a space, by which observations of the earth's surface (*Beobachtungen über den Erdraum*) are mentally ordered' (Reichert 1996: 19, translated by Judith Miggelbrink). According to this, mental maps are the outcome of 'ordering the world'.[6] A third understanding of mental maps that has some parallels to what is discussed by Istomin in this volume is closely related to an ongoing interest in 'how real human beings act and think in real settings, cultural and environmental'. Consistently, the focus has been shifted from mental maps as objects towards 'mapping behaviour' as (part of) practical doings. 'Cognitive mapping' is summarized as processes that enable people 'to comprehend and act effectively in the geographical environment' (Blaut et al. 2003: 166) and, therefore, understood as a universally human ability acquired by and required for 'geographical, macroenvironmental behaviour' (Blaut et al. 2003: 166). However, this 'does not prerequire any cognitive understanding of formal geometric principles in order to be effectuate' (Blaut et al. 2003: 177, see also Blaut 1999).

Although the significance of cognition and cognitive processes for spatial perception and behaviour has been acknowledged in general, geographical studies have usually not taken a closer look at mental operations. Nevertheless, processes in the brain have been assumed to directly precede actual and observable behaviour, or articulated preferences, aversions and images towards certain places. Models of cognitive mapping from the early 1960s to the late 1980s, compiled by Kitchin and Blades (2002: 14–23) in their overview on *The Cognition of Geographic Space*, include such elements as Pocock's model of 'basic physiological make-up' (Kitchin and Blades 2002: 16) that evoke an association of including mental operations in the *make up*, but do not go further into them. There are a few exceptions to the overall finding that cognition – though regularly mentioned – is widely neglected in models of spatial perception and cognitive mapping (Gold

6 In this context, 'individual' does not mean, as Reichert emphasizes, that the spatial logic of world ordering – even though represented through an individual – is entirely subjective, as it has to be understood as depending on 'the contextual network of meaning in a given culture' (Reichert 1996: 22, translated by Judith Miggelbrink). Furthermore, as these maps are drawn maps, they are part of communication processes and, therefore, based mainly on intersubjectively comprehensible modes of depiction.

1980). Occasionally, models from other disciplines focusing on cognitive processes (for example, Schmajuk and Thieme 1992) have recently been considered and applied to geographical research (Kitchin and Blades 2002). It may be supposed – and there is some evidence for this interpretation – that geography's abstinence from investigating processes in the brain results from the discipline's prevalence for macro-scale processes (Blaut 1999: 513) and subsequent methodological decisions. As a consequence, cognitive processes have not received the same attention as outward behaviour since these processes take place on a micro-scale thus requiring entirely different methodological approaches. A way forward may possibly be conducting detailed experiments in *the field* with a rather small sample of individuals but a wide documentation of their personal background and skills, and here anthropology could make its methodological contributions in the way sketched out above.

Spatial Production in Geography and Anthropology

In human geography as well as in other disciplines it has become a kind of truism that space is neither an ensemble of natural prerequisites for human activity nor a mere container in which human perception takes place. Instead, space has been subjected to intense theoretical reflections (for example, Crang and Thrift 2000, Massey 1993, 1994, 1999, Rose 1993, Simonsen 1996, Soja 1989, Weichhart 1993, 1999) thus being regarded as 'socially constructed' or 'socially produced'. Although both wordings are occasionally used interchangeably, they have different theoretical backgrounds. The notion of the socially produced nature of space – on which we focus here for the sake of the volume's intention – is deeply rooted in a critical approach towards the history and development of modern (Western, capitalist) society.

As concerns geography, methodological debates of the production of space as well as analyses of concrete appropriation of space through economic and political practices have initially taken place among radical and left-wing geographers (for example, Harvey 1990, Massey 1994, Smith 1984, Soja 1989, see also Belina and Michel 2007). There are at least two intellectual sources of the debate: First, ideas about the production of space entered geography through a reading of non-geographic writers such as Henri Lefebvre and Michel Foucault, who had already been concerned with the role of space in society. The work of Henri Lefebvre on *The Production of Space* (Lefebvre 2002 [1974]) has been used by several scholars, among them Andy Merrifield (2000), Christian Schmid (2005) and Neil Brenner and Stuart Elden (2001, 2009). Though it cannot be denied that productions of space have mainly been approached from a critique of capitalist society – thus having a very specific origin in terms of time and place – and even 'the wording … points at a highly specific cultural and historical context', as Stanek (2011: 81) states for the work of Lefebvre, it aims at offering a transcontextual conceptual language, a language of 'transcultural and transhistorical character' (Stanek 2011: 81), adequate to enable an understanding of spatial logics being at work in any part of the world.

Another intellectual source of reasoning is the work of Michel Foucault. His work on heterotopia (Foucault 1986) has been regarded as giving evidence to the fundamental meaning of space in the organization of society – what has been called the 'reassertion of space' (Soja 1989, for an extended discussion see Philo 1995, 2000). Recently, the focus of geographers has shifted slightly towards Foucault's history of governmentality (2007 [1977–78]) and a discussion of the role of space as a medium of the exertion of power. This part of Foucault's work has mainly been picked up by writers interested in questions of governmentality and statehood as well as transformations of the modern state (for example, Elden 2007a–c, 2010, Huxley 2006, Jessop 2007, Morton 2010, Rose-Redwood 2006) – even though 'the state' was not given very much attention by Foucault himself.

Notwithstanding the differences in how Lefebvre and Foucault understand the role of space in modern society and state, there are at least two general assumptions that link their work. First, by emphasizing the production of space it is argued that space should neither be understood as a pure and theoretical abstraction, nor as a mere stage or container holding social life, nor as an ensemble of natural and artificial prerequisites of human activity. Space does not precede the social but emerges from the organization of political and economic life and the conditions under which everyday life takes place in modern societies. Accordingly, the focus is not on an abstract theory of space but on a conceptualization of its social meaning and function as well as on the effects that a space, once it has been produced, may generate. Second, to understand space as socially produced does not only mean that all concrete spaces – spaces that people are confronted with in their everyday life such as borders (see Koch, this volume), public buildings, sacred places etc. – are socially produced. It also means that all common spatial concepts – including concepts of space as a place, space as a network, space as a homogenous abstraction, space as a container, lived spaces and mapped spaces (see Mazzullo, this volume) – are likewise socially produced (Belina and Michel 2007: 9). Therefore, any investigation of spatial dimensions of society has to be sensitive to the concrete appropriation of space as well as to how hegemonic concepts of space – such as the territory as the spatial form of the state – emerge and how they affect social life.

A second way of unfolding a geographic perspective on the production of space has been to enrich social theories towards an increasing sensitivity for space. Such a perspective has been offered by David Harvey who has been working on Marxist theory for several decades (Belina 2011: 241 *et passim*, Harvey 1990, 1993, 1999, 2000a, 2000b, 2001, 2005) and, in doing so, explicating the geographic dimension of Marx's analysis of capitalist society. Space is, to very briefly summarize his argument, necessarily produced by capitalist societies to overcome the inherent contradictions of capitalist production, namely to solve the problem of over-accumulation (see Belina 2011). Space, thus, is not a by-product of social life but a necessary tool to ensure its reproduction. Therefore, space is not only produced but strategically produced as a means to certain ends. These 'ends' are defined by capitalism itself thus integrating socio-spatial realities into circles of capitalist production and reproduction that are circles of investment and consumption.

As mentioned earlier, *anthropologists* began to problematize systematically concepts of space in the 1990s (Gupta and Ferguson 1992). Along with space, place and landscape[7] became used as key terms in many case studies and a few general works and collections (Bender 1993, Feld and Basso 1996, Hirsch and O'Hanlon 1995, Ingold 2000: 189–208, Jordan 2011, Ruotsala 2008). A succinct theoretical overview on space has recently been presented by Setha Low (2009). She too differentiates between *production* and *construction* of space, using the former in a more material and historic sense and defining the latter as 'the phenomenological and symbolic experience of space as mediated by social processes such as exchange, conflict, and control' (Low 2009: 24). She also connects seemingly disparate research themes – notably embodied space, on the one hand, and transnational/translocal spaces, on the other – to offer tools for analysing personal trajectories 'throughout the world' at the experiential level, while acknowledging that the number of possible pathways is limited by economic, political, and social constraints. Here we can see a strong link to scholarship on transnationalization (Glick Schiller and Faist 2010) and the even vaster literature on globalization, elaborately developed in anthropology (Appadurai 1996) and many other disciplines.

These themes have of course been explored in many settings and relations, yet when it comes to the North that lies beyond the 'global' North (Comaroff and Comaroff 2012: 45) we find that attempts at theorizing globalization and transnationalization are mysteriously scarce. The same is true for Siberia and also Mongolia. Studies on migration, settlers, long-distance commuting and transnational flows of people in the Russian Far North have been conducted very recently (see below), yet the conventional approach is to do ethnographic research within a predefined regional frame, thereby artificially insulating the community under study. Anthropologists thus participate to quite some degree in the (discursive) production and reproduction of nomadic and indigenous spaces. However, the 'classical' preference for detailed studies in rather remote communities provides a strong and unsurpassed methodological basis for studying indigenous place names, cosmologies and ways of narratively constructing space. These themes also relate to the unabatedly intensive research on different forms of traditional ecological knowledge – commonly abbreviated as TEK – and the issue of intellectual property that accompany the trend towards marketization of TEK (Krupnik and Vakhtin 1997, Nadasdy 1999, Ruotsala 2011, Wenzel 1999).

While few authors actually write about production of space explicitly, there is a wide range of anthropological studies – in the North and elsewhere – that illustrate and analyse how power differentials and hegemony shape spaces

7 As we found out as part of our literature review landscape became a popular concept in anthropology at about the same time it lost explanatory significance in geography.

(and consequently, the economic conditions for those living there). Of special interest is the process of designating exclusive territories and zones with specific regimes, for example nature reserves and other protected areas. Access to land and resources has been studied by many anthropologists in Siberia and the Far North (Anderson 2000, Fondahl 1998, Kasten 2002, Stammler 2005b) and recently been discussed from the perspective of environmental justice (Crate 2009, Donahoe 2009, Metzo 2009). A particular focus of this literature is the question of how state-induced modernization and reconstruction of nomadic and indigenous spaces has fundamentally changed systems of hunting and pastoralism (for example, Anderson 2000, Beach 2000, Golovnev and Osherenko 1999, Habeck 2005, Jordan 2003, Paine 1996, Stammler 2005a, Ventsel 2005, Vitebsky 2005, Vladimirova 2006, Wiget and Balalaeva 2011, Ziker 2002).

Nomadic and Indigenous

Strikingly, nomadic and indigenous groups have been the object of study for many generations of historians, geographers, and anthropologists alike. Let us look at the fascination with nomadism first, which goes back into ancient Greek times (see Prager 2012). Nomadic communities have always had close relationships with settled communities and sedentary political organizations (Khazanov 1984). Historically, there were manifold forms of interaction of nomadic and sedentary groups, oscillating between peaceful and violent, with power differentials on both sides. Sedentary individuals, groups and whole populations perceived and actually encountered nomads with changing degrees of fear, suspicion, incomprehension, condescension, and romanticism. Empires in Europe were swept away by nomadic invasions more than once. Nomads conquered most of Eurasia in the mid-thirteenth century, and in subsequent decades and centuries, dynasties of nomadic heritage acted as overlords of sedentary populations in China, Russia, Iran and Central Asia for decades or even centuries (Sneath and Kaplonski 2010, Weiers 1986). The confrontation, cooperation and mingling of sedentary and nomadic groups has thus created sustained interest in many academic disciplines. So much has been written about nomads by geographers[8] and anthropologists[9] that we cannot

8 Besides a large number of regional studies there is a number of general works in geography (for example, Scholz 1991, 1995, 1999 and Gertel 2002). A more popular-scientific overview was edited by Gertel and Calkins (2011).

9 Exemplary publications of the older period were presented by Salzman (1980) and Galaty and Salzman (1981). See Khazanov (1984) for a comprehensive and still authoritative overview, and Khazanov and Schlee (2012) for a recent description of the state of the art. A whole journal is devoted to *Nomadic Peoples*, and one of its special issues addresses reindeer pastoralism and nomadism in the Arctic in particular (Beach and Stammler 2006).

aspire to summarize the many strands of inquiry. What we do want to pursue here, though, is a juxtaposition of *nomadic* and *indigenous*.

With this in mind, let us now turn to the category of *indigenous*. Much younger in its etymology and usage, the term has seen widespread application in geography, anthropology, and political sciences since about the 1970s.[10] It has replaced earlier notions that for various reasons are now seen as deeply problematic, such as *tribal people* or *Naturvölker* (see above). More importantly, ethnic groups that see themselves as marginalized (by a majority population or colonial elite) and whose spokespeople fight for recognition and justice, have started to apply the term *indigenous* for self-designation. Since the mid twentieth century, a global indigenous movement has come into existence. Importantly, indigenous movements grew strong earlier in the Nordic countries and North America than in most other parts of the world (with the exception of Australia and New Zealand). Land claims by indigenous inhabitants in what is called the global South are met with much harder, often violent, responses on the part of the state. Many African states have acquired their independence comparatively recently, and the constitution and government do not acknowledge any claims of indigeneity; instead, they stress citizenship as a basis for equal rights (Niezen 2003: 69–76). By contrast, in the domestic colonies[11] of the Arctic, the ways of dealing with indigenous claims to land seem more placid and indigenous spokespeople proficiently use the state's rhetoric on welfare and recognition of minority rights for their cases.

Nomadic and *indigenous* denote, very obviously, two different categories. *Nomadic*, very roughly speaking, refers to the quality of a person of being at home in more than just one physical location. Nomadism is about moving. This is, at first sight at least, in contradiction to the notion of *indigenous*. Indigenous are those that were *there* before the advent of some other group or population. This quality is about being rooted in a place, and about not having moved. What may appear odd, then, is the observation that for most of our authors, nomadic and indigenous *are* very much the same. This deserves some commentary. Most of the chapters in this volume draw on research in the Far North – the Arctic and sub-Arctic – and tell about the livelihoods of groups that, in a region of extreme changes in the availability and abundance of plants and prey, simply had to be physically mobile and flexible in their actions. We may consider mobility a virtue of life in the Far North (Vitebsky 2005).

The arrival (read: encroachment) of people moving from the zone of moderate climate into the North did not simply reduce the mobility of the indigenous groups

10 Brown and Sant (1999), Friedman (2008), Niezen (2003), Zenker (2011) provide useful introductions into the concept of indigeneity. For Russia, see Donahoe et al. (2008), Sokolovski (2007).

11 We claim that the Arctic has not been de-colonized, with Greenland and Nunavut being those regions that have attained the highest degree of autonomy. 'Domestic' refers to the fact that there is no clear divide between centre and colony, as is the case with overseas colonies.

they encountered: in a number of cases, the opposite is true.[12] However, since the times of the Enlightenment and with the advent of missionaries of various projects of social engineering, mobility itself became a problem: rulers and their bureaucratic agents were concerned about the unpredictability and fuzziness of the whereabouts of their subjects – hence the need to draw boundaries – and missionaries of various religious and revolutionary denominations were concerned about the uncivilized and wild nature of nomads, whom they perceived as standing on a lower stage in the grand evolutionary scheme of human kind – hence the desire to make them sedentary (Slezkine 1994). Anthropology's concern with nomadic and indigenous groups initially emerged from these premises too.[13] Importantly, however, the majority of anthropologists have seen themselves as advocates of indigenous peoples, in earlier centuries as much as now, trying to sketch out the logic of the Other and explicating how earlier, present, or future encroachments exert their impact on indigenous ways of being. Key to these explanations is 'the land'. Since nomadic and indigenous livelihoods depend on biological resources, and since they depend on knowing and living with the land, its animals and its spirits, land is always good in the perception of anthropologists.[14] Hence space is not seen as a problem but as an asset. Seemingly, space only *becomes* a problem when outsiders (settlers, bureaucrats, entrepreneurs etc.) try to encroach on land, when they try to draw boundaries, single out parcels of land as commodities to be traded for exclusive use of some individuals or industries.

Space: Striated or Smooth or Both?

The latter observation – about the partitioning of nomadic space – seems to be in line with the binary concept of space proposed by Gilles Deleuze and Felix Guattari,[15] which has received significant attention in many disciplines (in

12 As the Russian Crown required nomads to pay tribute in form of pelts, hunters often extended their territory of hunting to fulfil the officials' hunger for furs. To give one more example, the development of large-scale, market-oriented reindeer husbandry and the concomitant extension of migration routes can also be ascribed to economic changes induced by colonial policies of the Russian Empire (Krupnik 1993: 160–71, Sasaki 2010) and the Scandinavian kingdoms (Pennanen and Näkkäläjärvi 2003).

13 Vermeulen (2011) claims that the very discipline of *Ethnologia* emerged from eighteenth-century explorers' studies of non-Russian peoples in Siberia.

14 For a critical reading of the inherently good quality of 'land', see Rethmann (2004: 87, 93).

15 Deleuze and Guattari introduce and discuss *l'espace lissé et l'espace strié* (smooth and striated space) in the chapter '1227 – Traité de Nomadologie: la machine de guerre' of their book *Mille Plateaux* (1980: 447 and 471–482), translated as '1227: Treatise on Nomadology – the War Machine' in *A Thousand Plateaus* (1987: 361–2 and 380–87). In a subsequent chapter ('1440 – Le lissé et le strié') they expand on this distinction (1980: 592–625, 1987: 474–500).

geography, for example, Doel 2000, Matless 1997, see also Casey 1997: 301–9). Deleuze and Guattari distinguish between smooth and striated space. Smooth space – synonymously called 'nomad space' (1987: 384) – is characterized as 'vectorial' (1987: 361), as an almost infinite, ever-changing and fractal-like repetition of its own surface (notably the sea and the steppe, by logical extension also the tundra and the polar ice shields); whereas striated space is 'metric' (1987: 362), divided and delineated. Striated space must be marked by land surveyors because it is under the surveillance of the (settlers') state. 'One of the fundamental tasks of the state is to striate the space over which it reigns, or to utilize smooth spaces as a means of communication in the service of striated space' (1987: 385). What lies in between striated spaces is smooth space, and Deleuze and Guattari associate smooth or nomad space with the acts of rebellion, insubordination and guerilla warfare (1987: 385). Several of the contributors to this volume borrow Deleuze's and Guattari's binary model of space to theorize conflicts over land, when flexible arrangements of land use (for example in the case of hunting and herding) are contested by the zeal for partitioning on the part of state institutions and resource-extracting industries.

Where we differ with Deleuze and Guattari, however, is the ideal type of *nomad* they construct. The contributions to this volume do not rely on such an idealized (and indeed romanticized) notion of the nomad; instead, we derive our insights from our encounters and extended stays with people who regard themselves as nomads and pastoralists. From our variegated experience, we can claim that relations between nomads and state (bureaucrats) comprise far more than rebellion and warfare. Solidarity, cooperation, complicity, make-do, and multiple hidden and open forms of resistance all form part of a wide range of responses and proactive strategies and policies that nomads and indigenous groups employ in their dealings with sedentary or external groups and their institutions. In other words, nomads and indigenous individuals do not lead a liminal existence in between striated spaces. We think that the spatial concept proposed by Deleuze and Guattari requires some refinement to become meaningful in more than just a philosophical sense. Of each place, we believe, it may be said that it has striated and smooth qualities. Spatial cognition depends on memories, names, landmarks, and other marks and inscriptions; simultaneously it depends on the perception of the continually changing weather, texture of the snow, vegetation, and other fractal-like patterns and phenomena of the landscape. And the very process of spatial production – each act of making a space – amounts to marking that space, inscribing it with functions, meanings, and ultimately also hierarchies.

Weaving the Threads Together

Having outlined the disciplinary background of the authors and editors, their two-pronged theoretical approach to space, and our contextualization of nomadic and indigenous, we can now give a brief outlook of the topics where the disciplines, approaches, and distinct bodies of scholarship reconverge and provide fruitful ground for further inquiry. Regarding the empirical work conducted within the study of nomadism and indigeneity numerous spatial aspects have been addressed; among them, four partly intertwined strands have recently gained particular attention and are promising themes for future research:

1. First, a range of studies has addressed space in terms of land use, access to land, struggles for control over and protection of land and pastures, regulation and management of land and natural resources. We have listed Siberian examples, here we add case studies from Finland (Belina and Miggelbrink 2012, Miggelbrink and Mazzullo 2011) and from Inner and High Asia (Janzen 2011, Kreutzmann 2009, 2012a, 2012b, Sneath 2004).[16] While these questions of control over land have often been described and discussed under the headline of production of space (Heikkilä 2006, Koch and Miggelbrink 2011, Raitio 2008, Yiftachel 2006), the cognitive aspects of changes in land use and resource management have been described less frequently, and we need to ask why. After all, different modes and regimes of land use rely on particular skills, specific assessments of the *values* or affordances that the land has to offer, and elaborate symbolic systems which inform, and are informed by, spatial cognition. Among the few studies that take an explicitly cognitive approach to conflicts over land and resource use in a northern setting is that of Atran and Medin (2008: 225–54).
2. A second strand of inquiry puts the focus on space in terms of territory and territoriality by examining the incorporation of the so-called *terra nullius* (commons, barren grounds, *wilderness*, moors and marshes etc.) into emerging nation states and administrative systems. Several scholars have recently observed and analyzed the nation states' zeal for mapping and demarcating spaces, usually to the detriment of nomadic and indigenous economies and sovereignties, and often accompanied by changes in local people's perceptions of the land they inhabit (Anderson 2000, Fondahl 2005, Karppi 2001). This is not to say, however, that states and indigenous people are consistently treated as antagonist actors; rather, the increasing autonomy of indigenous groups and their equally increasing capacity to act politically in different arenas beyond the nation-state call for new perspectives on networks and assemblages through which indigenous interests are articulated.
3. A third strand comprises studies on indigenous classifications of places and landscapes, the meaning of place, such as place-making and place-naming, the importance of places and place names in people's biographies and, by the same token, the imprint and recognition of people's deeds, mishaps,

16 This also comprises the question of large-scale purchases of communal lands that have come to be discussed under 'land grab', most prominently in African states (Gertel and Le Heron 2011, see also Pearce 2012).

and misdeeds in place names (Castree 2004, Larsen and Johnson 2012, and in northern settings, for example, Collignon 2006a, 2006b, Cruikshank 2005, Istomin 2012, Kari and Fall 2003, Mazzullo 2009, Thornton 2008, Ulturgasheva 2012).

4. Finally, the last five to ten years have seen novel research on mobility and movement, including long-distance commuting (Stammler and Eilmsteiner-Saxinger 2010), migration in the Far North (Heleniak 2010, Huskey and Southcott 2010), settler populations (Thompson 2008) and phenomenological studies of travelling in the North (Argounova-Low 2012, Habeck 2006, Van den Scott 2009). There is also growing research on the organization of pastoralists' movements via GPS, changing conditions of travelling owing to the use of mobile phones and other technologies, changing methods of orientation, and new methods to describe, depict, and comprehend indigenous and nomadic spaces (Aporta and Higgs 2005, Stammler 2009).

While by no means exhaustive,[17] this short outline serves to identify common ground for transdisciplinary research. What is more, it is triggered by and feeds back into research agendas of indigenous peoples' organizations themselves. While for many years, scientists and non-indigenous researchers have tended to treat the Far North as a pristine, almost unimpacted *terra nullius* and largely neglected the manifold material and symbolic interactions of local populations with the land, they are now demanded and encouraged to seriously engage with emic concepts and systems of knowledge that divert from academic conventions (for example, Anderson and Nuttall 2004).

Contributions to the Volume

The four strands outlined above also provide the basis for this volume's contributions. Most of them build on 'Far Northern' case studies to theorize nomadic and indigenous spaces, whereas two chapters and the epilogue apply a general scope. We finish the introduction with a short overview of the individual chapters.

Denis Wood challenges central axioms of geography by introducing the concept of places without location. He discusses two opposite examples of being in the world, one of a frequent traveller and one of a person who is deeply rooted in one place. Despite these differences both have a permanent address, as bureaucracies prefer people to be rooted in a location, whereas 'people in motion are a threat'. As maps enable control of the land, people without an address are not accommodated

17 To give an example of the thematic fields that we cannot cover here: virtual space, cyberspace, and internet-based social networking sites are of growing importance for indigenous communities and individuals throughout the Far North.

by the state. In order to resolve conflicts over land tenure for those who have no fixed addresses, such as nomads, travellers, and refugees, Wood pledges to 'guarantee a person's *place in the world* against a minimal right to a *location on a map*'. Equipped with such an entitlement, people who do not have a territory or a fixed residency could then access services and land when they need it.

In a similar vein, Denis Retaillé challenges the 'sedentary paradigm' among scholars of geography by proposing a new mobile space theory. The 'mobile

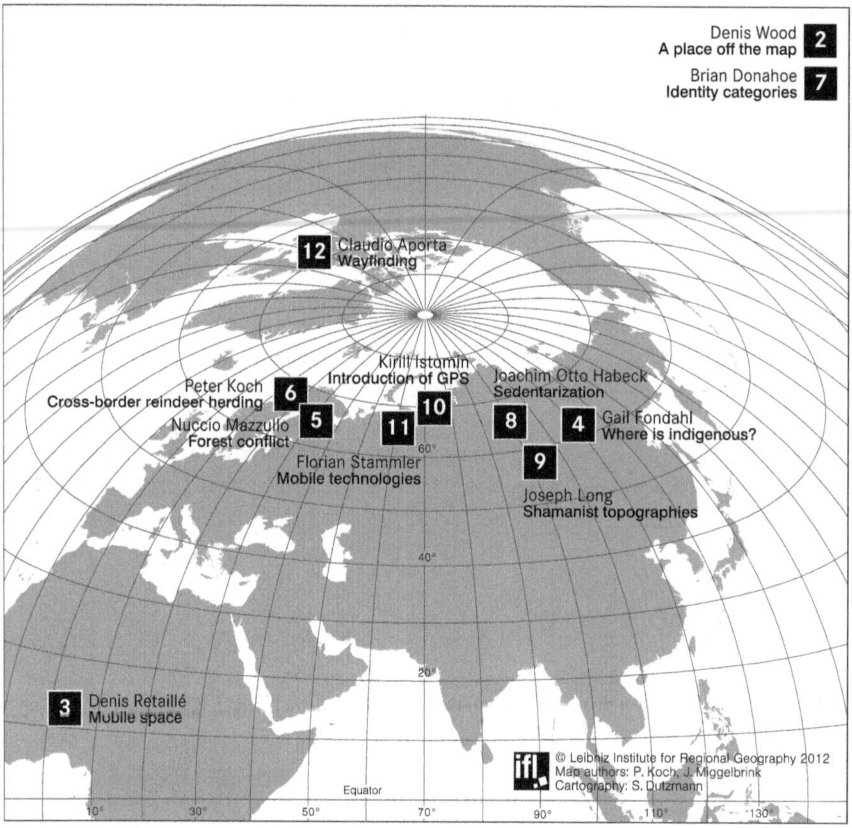

Map 1.1 Regional setting of chapters

space paradigm' accommodates the idea of movement being constitutive of place. Retaillé's account of a village in East Niger shows how a place is not bound to a location on a map, but is itself mobile. The village is recognized as one and the same by its visitors and inhabitants, even though it has moved to a new location, because of the 'co-presence' of its inhabitants. The paradigm suggested by Retaillé enables the analysis of people's movement in terms of scattering (dispersion), gathering (polarization) and exchange (crisscrossing of flows), and thus the analysis of people's migration in a globalized world.

The chapter by Gail Fondahl, starting from a relational notion of space, describes the involvement of the state in the production of space. In particular, she shows how the production of discrete spaces through different legislations in the first two decades of the Russian Federation fails to account for indigenous spatiality. By defining 'where is indigenous', rather than 'who is indigenous', the Russian state partitions its territory, thus ignoring 'indigenous lived experience and spatial practices' both of urban dwellers and traditional land users. In particular, the state privileges a specific notion of traditional activities, which is then spatialized.

In the case of a conflict over the use of natural resources in Finnish Lapland, Nuccio Mazzullo shows how official redefinition of traditional reindeer herding pastures as commercial forests has produced a conflict lasting already over two decades. Moreover, the state is in a contradictory position, for it is obliged to protect indigenous traditional livelihoods and forest biodiversity, while at the same time it must make profits by extracting natural resources from areas where Sámi practice reindeer herding. Maps play a crucial role in the social production of space. Nevertheless, Mazzullo argues, it was only when the Sámi herders strategically used the notion of indigeneity that they were able to reach an agreement on a moratorium on forest felling in their pastures.

The Norwegian and Swedish states similarly disrupt the traditional livelihood of Sámi reindeer herders, as Peter Koch highlights in his chapter. Although the states reserve exclusive rights to indigenous people, they fail to fully acknowledge the indigenous production of space, which requires crossing state borders between Norway and Sweden. Furthermore, differences in the legislation of the two countries allow unequal access to resources of some Sámi reindeer herders over the rest, creating a conflict that is fought on the basis of citizenship, rather than indigenous status. Drawing on Foucault, Koch describes different mechanisms of power at play – sovereignty, discipline and security – and shows how these technologies are fundamental for the states to control their territories.

Reflecting on categories of anthropological study, Brian Donahoe challenges these approaches for not taking into account the insights provided by brain or cognitive sciences. Critically examining the work of various social scientists, he identifies a lack of scholarly engagement with processes of socio-cognitive standardization, which entail 'the mutually constituting dynamics between brain, body, and the environment'. One of the examples Donahoe refers to is the use of new terms by groups of people to describe themselves. He argues that the linguistic and conceptual implications for the cognitive internalization of new categories should be taken more serious in the anthropological study of indigenous peoples.

Joachim Otto Habeck argues in his chapter that the administrative restructuring in Central Siberia, as part of the Soviet state induced process of indigenous people's sedentarization, had an impact on spatial cognition among the indigenous Evenki reindeer herding nomads. Starting from Fondahl's concept of 'socialist land enclosure' (1998), Habeck introduces the concept of 'cognitive enclosure' to explain the effects of sedentarization as well as the introduction of new technologies on people's perception of space and on the skills pertaining to movement. To illustrate these changes Habeck describes processes of learning and forgetting regarding, for instance, ways of living in new homes and associated body postures as well as modes of travelling and methods of wayfinding.

Joseph Long describes how a shamanist topography has long existed outside official forms of national culture in Cisbaikalia in southern Siberia. Long highlights the recent development whereby Buryat shamanists are bringing their topography into public view, with the implication that sacred places are being 'syncretized with discourses of cultural heritage and conservation along with a growing awareness of tourism'. He emphasizes that Buryats assert their sense of belonging to the land in public discourse. This process involved mapping sacred places. Here, Long highlights the notion of place, constituting meaningful phenomena in human experience, versus the notion of neutral space in administrative territories.

Continuing the discussion on the influence of GPS on the cognitive perception of space among Nenets reindeer herders, Kirill Istomin presents the case of nomads in the Taz Tundra, north-western Siberia. Old methods of wayfinding are increasingly becoming obsolete among the young generation of herders. Istomin highlights three consecutive stages, in which new technology is initially used complementarily, then increasingly in parallel and finally in substitution of traditional methods. Where formerly way-finding required sensing the direction of the wind, ground structure and time passed, GPS technology has replaced these skills. His theoretical aim is to extend the field of analysis to other types of innovation.

How technologies are creatively integrated into the flow of movement and communication among reindeer herders of Yamal Peninsula, north-western Siberia, is accounted for by Florian Stammler. He describes how Nenets see particular affordances in innovations, and in this analysis he focuses on rubber boots, snowmobiles, mobile phones and GPS. The diffusion of technologies among nomadic societies presumes an 'interplay of cognitive activity, experience and openness to introducing new items to their own culture'. Stammler points out that the adoption of new technologies creates new spaces of communication, but at the same time does not reduce people's experience-based relations to the environment.

Claudio Aporta offers an ethnographic account of how spatial categories provided by new technologies effect travelling and way-finding among the Inuit. GPS is considered an opportunity, but not adapted uncritically, as the community also discusses potential dangers. Aporta shows how Inuit hunters, who formerly relied on their intimate knowledge of the landscape, are now increasingly affected by the 'power of the abstract principles of GIS', which initially required an immersion into Western cartography. Inuit hunters have become part of an ecological system that connects GPS users with a structure of interdependent technologies and a whole network of social organizations.

In the epilogue, Tim Ingold rounds off the volume, returning to the question of the role of cognition in environmental perception and movement. He claims that neuro-imaging or brain-mapping cannot fully account for the dynamics of human existence. Similar to maps, these techniques of representation depict structures along which activity or movement did, may and will occur, but not the activity or movement itself. He also emphasizes that to be able to fully appreciate the rhythm of life – in nomadic as much as in sedentary contexts – we must pay attention not only to movement but also to rest. His words are a reminder of the basic qualities of life, as coming and dwelling and going; and over all our musings about space, we should not forget the importance of temporality in the experience of space.

References

Agnew, J. 1997. Representing space: Space, scale and culture in social science, in *Place/Culture/Representation*, edited by J. Duncan and D. Ley. London: Routledge, 251–71.

Anderson, D.G. 1991. Turning hunters into herders: A critical examination of Soviet development policy among the Evenki of southeastern Siberia. *Arctic*, 44(1), 12–22.

Anderson, D.G. 2000. *Identity and Ecology in Arctic Siberia: The Number One Reindeer Brigade*. Oxford: Oxford University Press.

Anderson, D.G. and Nuttall, M. (eds). 2004. *Cultivating Arctic Landscapes: Knowing and Managing Animals in the Circumpolar North*. New York: Berghahn.

Anisimov, A.F. 1963. Cosmological concepts of the Peoples of the North, in *Studies in Siberian Shamanism*, edited by H.N. Michael. Toronto: University of Toronto Press, 157–229.

Antonsich, M. 2010. Rethinking territory. *Progress in Human Geography*, 35(3), 422–5.

Aporta, C. 2004. Routes, trails and tracks: Trail breaking among the Inuit of Igloolik. *Etudes/Inuit/Studies*, 28(2), 9–38.

Aporta, C. and Higgs, E. 2005. Satellite culture: Global positioning systems, Inuit wayfinding, and the need for a new account of technology. *Current Anthropology*, 46(5), 729–53.

Appadurai, A. 1996. *Modernity at Large: Cultural Dimensions of Globalization*. Minneapolis, MN: University of Minnesota Press.

Argounova-Low, T. 2012. Narrating the road. *Landscape Research*, 37(2), 191–206.

Atran, S. and Medin, D. 2008. *The Native Mind and the Cultural Construction of Nature*. Cambridge, MA: MIT Press.

Barnett, C. 1998. The cultural turn: Fashion or progress in human geography? *Antipode*, 30(4), 379–94.

Beach, H. 2000. Reindeer-pastoralism politics in Sweden: Protecting the environment and designing the herder, in *Negotiating Nature: Culture, Power and Environmental Argument*, edited by A. Hornborg and G. Pálsson. Lund: Lund University Press, 179–211.

Beach, H. and Stammler, F. 2006. Human–animal relations in pastoralism. *Nomadic Peoples*, 10(2), 6–30.

Belina, B. 2011. Kapitalistische Raumproduktionen und ökonomische Krise: Zum Begriff des spatial fix bei David Harvey [Capitalist productions of space and economic crisis: On David Harvey's notion of spatial fix]. *Zeitschrift für Wirtschaftsgeographie*, 55(4), 239–52.

Belina, B. and Michel, B. 2007. Raumproduktionen: Zu diesem Band [Productions of space: On this volume], in *Raumproduktionen: Beiträge der Radical Geography – eine Zwischenbilanz*, edited by B. Belina and B. Michel. Münster: Westfälisches Dampfboot, 7–34.

Belina, B. and Miggelbrink, J. 2012. Raum, Recht und Indigenität – Zu den Kämpfen um Landrechte indigener Völker am Beispiel der Sámi in Finland [Space, law, and indigeneity: On indigenous peoples' rights to land, with reference to the Sámi of Finland]. *Peripherie*, 126/127, 190–217.

Bender, B. (ed.) 1993. *Landscape: Politics and Perspectives*. Providence: Berg.

Berger, T.R. 1985. *Village Journey: The Report of the Alaska Native Review Commission*. New York: Hill & Wang.

Blaut, J.M. 1999. Maps and spaces. *Professional Geographer*, 51(4), 510–15.

Blaut, J.M., Stea, D., Spencer, C. and Blades, M. 2003. Mapping as a cultural and cognitive universal. *Annals of the Association of American Geographers*, 93(1), 165–85.

Brenner, N. and Elden, S. 2001. Henri Lefebvre in contexts: An introduction. *Antipode*, 33(5), 763–68.

Brenner, N. and Elden, S. 2009. Henri Lefebvre on state, space, territory. *International Political Sociology*, 3(4), 353–77.

Brown, J.N. and Sant, P.M. (eds). 1999. *Indigeneity: Construction and Re/Presentation*. Commack, NY: Nova Science Publishers.

Carpenter, E. 1973. *Eskimo Realities*. New York: Holt, Rinehart & Wilson.

Casey, E. 1997. *The Fate of Place: A Philosophical History*. Berkeley, CA: University of California Press.

Castree, N. 1999. Situating cultural twists and turns. *Environment and Planning D: Society and Space*, 17(3), 257–60.

Castree, N. 2004. Differential geographies: Place, indigenous rights and 'local' resources. *Political Geography*, 23(2), 133–67.

Chorley, R.J. and Haggett, P. (eds). 1967. *Socio-Economic Models in Geography*. London: Methuen.

Collignon, B. 2006a. Inuit place names and sense of place, in *Critical Inuit Studies: An Anthology of Contemporary Arctic Ethnography*, edited by P. Stern and L. Stevenson. Lincoln, NA: University of Nebraska Press, 187–205.

Collignon, B. 2006b [1996]. *Knowing Places: The Inuinnait, Landscapes, and the Environment*, translated and edited by L. Weber Müller-Wille. [Edmonton]: CCI Press.

Comaroff, J. and Comaroff, J.L. 2012. *Theory from the South, or, How Euro-America is Evolving toward Africa*. Boulder, CO: Paradigm.

Cook, I., Crouch, D., Naylor, S. and Ryan, J.R. (eds). 2000. *Cultural Turns/ Geographical Turns: Perspectives on Cultural Geography*. Harlow: Prentice Hall.
Cosgrove, D. 1983. Towards a radical cultural geography: Problems of theory. *Antipode*, 15, 1–11.
Cox, K.R. 1972. *Man, Location, and Behavior: An Introduction to Human Geography*. New York: John Wiley & Sons.
Cox, K.R. and Golledge, R.G. 1981. *Behavioral Problems in Geography Revisited*. New York: Routledge.
Crang, M. and Thrift, N. (eds). 2000. *Thinking Space*. London: Routledge.
Crate, S. 2009. Viliui Sakha of subarctic Russia and their struggle for environmental justice, in *Environmental Justice in the Former Soviet Union*, edited by J. Agyeman and Y. Ogneva-Himmelberger. Boston, MA: MIT Press, 189–213.
Cruikshank, J. 2005. *Do Glaciers Listen? Local Knowledge, Colonial Encounters, and Social Imagination*. Vancouver: UBC Press.
Deleuze, G. and Guattari, F. 1980. *Mille plateaux*. Paris: Editions de Minuit.
Deleuze, G. and Guattari, F. 1987. *A Thousand Plateaus: Capitalism and Schizophrenia*, translated by B. Massumi. London: The Athlone Press Ltd.
Doel, M.A. 2000. Un-glunking geography: Spatial science after Dr Seuss and Gilles Deleuze, in *Thinking Space*, edited by M. Crang and N. Thrift, London: Routledge, 117–35.
Donahoe, B. 2009. The law as a source of environmental justice in the Russian Federation, in *Environmental Justice in the Former Soviet Union*, edited by J. Agyeman and Y. Ogneva-Himmelberger. Boston, MA: MIT Press, 21–45.
Donahoe, B., Habeck J.O., Halemba, A. and Sántha, I. 2008. Size and place in the construction of indigeneity in the Russian Federation. *Current Anthropology*, 49(6), 993–1020.
Downs, R.M. and Stea, D. 1973. *Image and Environment: Cognitive Mapping and Spatial Behavior*. Chicago, IL: Aldine.
Elden, S. 2007a. Governmentality, calculation, territory. *Environment and Planning D: Society and Space*, 25(3), 562–80.
Elden, S. 2007b. Rethinking governmentality. *Political Geography*, 26, 29–33.
Elden, S. 2007c. Terror and Territory. *Antipode*, 39(5), 821–45.
Elden, S. 2010. Land, terrain, territory. *Progress in Human Geography*, 34(6), 799–817.
Feit, H.A. 2000. The Cree of James Bay, Quebec, Canada, in *Endangered Peoples of the Arctic: Struggles to Survive and Thrive*, edited by M.M.R. Freeman. Westport, CT: Greenwood, 38–57.
Feld, S. and Basso, K. (eds). 1996. *Senses of Place*. Santa Fe, NM: School of American Research Press.
Fondahl, G. 1998. *Gaining Ground? Evenkis, Land and Reform in Southeastern Siberia*. Boston, MA: Allyn & Bacon.
Fondahl, G. 2005. 'Everything is as if beyond a boundary': Reflections on apprehensions regarding aboriginal re-territorialization in northern Russia, in

Rebuilding Identities: Pathways to Reform in Post-Soviet Siberia, edited by E. Kasten. Berlin: Reimer, 89–107.

Foucault, M. 1986. Of other spaces. *Diacritics*, 16(1), 22–7.

Foucault, M. 2007 [1977–78]. *Security, Territory, Population. Lectures at the Collège de France 1977–78*, edited by M. Senellart. Basingstoke: Palgrave Macmillan.

Friedman, J. 2008. Indigeneity: Anthropological notes on a historical variable, in *Indigenous Peoples: Self-Determination, Knowledge, Identity*, edited by H. Minde, H. Gaski, S. Jentoft and G. Midré. Delft: Eburon, 29–48.

Galaty, J.G. and Salzman, P.C. (eds). 1981. *Change and Development in Nomadic and Pastoral Societies*. Leiden: Brill.

Gell, A. 1985. How to read a map: Remarks on the practical logic of navigation. *Man: The Journal of the Royal Anthropological Institute*, New Series, 20(2), 271–86.

Gertel, J. 2002. Globalisierung, Entankerung und Mobilität. Analytische Perspektiven einer gegenwartsbezogenen geographischen Nomadismusforschung [Globalization, uprooting and mobility: Analytical prospects of a present-day geographical research on nomadism], in *Nomadismus aus der Perspektive der Begrifflichkeit*, edited by S. Leder and B. Streck, Beiträge der 1. Tagung am 11.7.2001, Orientwissenschaftliche Hefte 3, Mitteilungen des SFB 'Differenz und Integration' 1, 57–88.

Gertel, J. and Calkins, S. (eds). 2011. *Nomaden in unserer Welt – die Vorreiter der Globalisierung: Von Mobilität und Handel, Herrschaft und Widerstand* [Nomads in our world – the vanguards of globalization: On mobility and trade, domination and resistance]. Bielefeld: Transcript.

Gertel, J. and Le Heron, R. (eds). 2011. *Economic Spaces of Pastoral Production and Commodity Systems: Markets and Livelihoods*. Farnham: Ashgate.

Gibson, K. 2001. Regional subjection and becoming. *Environment and Planning D: Society and Space*, 19, 639–67.

Gladwin, T. 1964. Culture and logical process, in *Explorations in Cultural Anthropology*, edited by W. Goodenough. New York: McGraw-Hill, 167–77.

Gladwin, T. 1970. *East is a Big Bird: Navigation and Logic on Puluwat Atoll*. Cambridge, MA: Harvard University Press.

Glick Schiller, N. and Faist, T. (eds). 2010. *Migration, Development, and Transnationalization: A Critical Stance*. New York: Berghahn.

Gold, J.R. 1980. *An Introduction to Behavioural Geography*. Oxford: University Press.

Golledge, R. (ed.). 1999. *Wayfinding Behavior: Cognitive Mapping and Other Spatial Processes*. Baltimore, MD: Johns Hopkins University Press.

Golledge, R. 2000. Cognitive maps, spatial abilities, and human wayfinding. *Geographical Review of Japan B*, 73(2), 93–104.

Golovnev, A.V. and Osherenko, G. 1999. *Siberian Survival: The Nenets and Their Story*. Ithaca, NY: Cornell University Press.

Gould, P. 1972. On mental maps, in *Man, Space, and Environment: Concepts in Contemporary Human Geography*, edited by P.W. English and R.C. Mayfield. Oxford: University Press, 260–82.

Gould, P. and White, R. 1974. *Mental Maps*. New York: Routledge.

Gregory, D. 1998. *Geographical Imaginations*. Cambridge: Blackwell.

Gupta, A. and Ferguson, J. 1992. Beyond 'culture': Space, identity, and the politics of difference. *Cultural Anthropology*, 7(1), 6–23.

Habeck, J.O. 2005. *What it Means to Be a Herdsman: The Practice and Image of Reindeer Husbandry among the Komi of Northern Russia*. Münster: LIT.

Habeck, J.O. 2006. Experience, movement and mobility: Komi reindeer herders' perception of the environment. *Nomadic Peoples*, 10(2), 123–41.

Haggett, P. 1972. *Geography: A Modern Synthesis*. New York: Harper & Row.

Harva [Holmberg], U. 1938. Die religiösen Vorstellungen der altaischen Völker [The religious ideas of Altaian peoples]. *Folklore Fellows Communications*, 125.

Harvey, D. 1969. *Explanation in Geography*. New York: Edward Arnold.

Harvey, D. 1973. *Social Justice and the City*. Baltimore, MA: Johns Hopkins University Press.

Harvey, D. 1990. Between space and time: Reflections on geographical imaginations. *Annals of the Association of American Geographers*, 80(3), 418–34.

Harvey, D. 1993. From space to place and back again: Reflections on the condition of postmodernity, in *Mapping the Futures*, edited by J. Bird, B. Curtis, G. Robertson and L. Tickner. London: Routledge, 3–29.

Harvey, D. 1999. *The Limits to Capital*. London and New York: Verso.

Harvey, D. 2000a [1989]. *The Condition of Postmodernity: An Enquiry into the Origins of Cultural Change*. 18th Edition. Malden, MA: Blackwell.

Harvey, D. 2000b. *Spaces of Hope*. Berkeley, CA: University of California Press.

Harvey, D. (ed.). 2001. *Spaces of Capital: Towards a Critical Geography*. Edinburgh: Edinburgh University Press.

Harvey, D. 2005. *Spaces of Neoliberalization: Towards a Theory of Uneven Geographical Development*. Stuttgart: Franz Steiner.

Heikkilä, L. 2006. *Reindeer Talk: Sámi Reindeer Herding and Nature Management*. Rovaniemi: Acta Universitatis Lapponiensis.

Heikkilä, K. and Fondahl, G. 2012. Co-managed research: Non-indigenous thoughts on an indigenous toponymy project in northern British Columbia. *Journal of Cultural Geography*, 29(1), 61–86.

Heleniak, T. 2010. Population change in the periphery: Changing migration patterns in the Russian North. *Sibirica*, 9(3), 9–40.

Hirsch, E. and O'Hanlon, M. (eds). 1995. *The Anthropology of Landscape: Perspectives on Place and Space*. Oxford: Oxford University Press.

Huskey, L. and Southcott, C. (eds). 2010. *Migration in the Circumpolar North: Issues and Contexts*. Edmonton: Canadian Circumpolar Institute Press, University of Alberta.

Hutchins, E. 1995. *Cognition in the Wild*. Cambridge, MA: MIT Press.

Huxley, M. 2006. Spatial rationalities: Order, environment, evolution and government. *Social and Cultural Geography*, 7(5), 771–87.

Ingold, T. 2000. *The Perception of the Environment: Essays in Livelihood, Dwelling and Skill*. London: Routledge.

Istomin, K. 2012. The land to herd and the space to travel: Comparing the categorizations of landscape among Komi and Nenets reindeer herding nomads, in *Nomadismus in der Alten Welt*, edited by L. Prager. Münster: LIT, 232–55.

Istomin, K. and Dwyer, M. 2009. Finding the way: A critical discussion of anthropological theories of human spatial orientation with reference to reindeer herders of northeastern Europe and western Siberia. *Current Anthropology*, 50(1), 29–49.

Janzen, J. 2011. Mongolian pastoral economy and its integration into the world market under socialist and post-socialist conditions, in *Economic Spaces of Pastoral Production and Commodity Systems*, edited by J. Gertel and R. Le Heron. Farnham: Ashgate, 195–209.

Jessop, B. 2007. From micro-powers to governmentality: Foucault's work on statehood, state formation, statecraft and state power. *Political Geography*, 26, 34–40.

Jordan, P. 2003. *Material Culture and Sacred Landscape: The Anthropology of the Siberian Khanty*. Walnut Creek, CA: Alta Mira Press.

Jordan, P. (ed.). 2011. *Landscape and Culture in Northern Eurasia*. Walnut Creek, CA: Left Coast Press.

Kari, J. and Fall, J. 2003. *Shem Pete's Alaska: The Territory of the Upper Cook Inlet Dena'ina*. Fairbanks: University of Alaska Press.

Karppi, K. 2001. Encountering different territorialities: Political fragmentation of the Sami homeland. *Tijdschrift voor Economische en Sociale Geografie*, 92(4), 394–404.

Kasten, E. (ed.). 2002. *People and the Land: Pathways to Reform in Post-Soviet Siberia*. Berlin: Reimer.

Khazanov, A. 1984. *Nomads and the Outside World*. Cambridge: Cambridge University Press.

Khazanov, A. and Schlee, G. (eds). 2012. *Who Owns the Stock? Collective and Multiple Forms of Property in Animals*. New York: Berghahn.

Kitchin, R. and Blades, M. 2002. *The Cognition of Geographic Space*. London: I.B. Tauris.

Koch, P. and Miggelbrink, J. 2011. Being in the frontline of a Sámi culture and a private business: Cross-border reindeer herding in northern Norway and Sweden. *Nomadic Peoples*, 15(1), 114–43.

Kreutzmann, H. 2009. *Three Years After: Evaluation of the GRC-ICRC Livestock Programme in the Earthquake-Affected Areas of Pakistan-Administered Kashmir*. Berlin: Centre for Development Studies (ZELF).

Kreutzmann, H. 2012a. After the flood: Mobility as an adaption strategy in high mountain oases – the case of Pasu in Gojal, Hunza Valley, Karakoram. *Die Erde*, 143 (1–2), 49–73.

Kreutzmann, H. (ed.). 2012b. *Pastoral Practices in High Asia: Agency of 'Development' Effected by Modernisation, Resettlement and Transformation*. Dordrecht: Springer.

Krupnik, I. 1993. *Arctic Adaptations: Native Whalers and Reindeer Herders of Northern Eurasia*. Expanded English edition, translated and edited by M. Levenson. Hanover, NH: University Press of New England.

Krupnik, I. and Vakhtin, N. 1997. Indigenous knowledge in modern culture: Siberian Yupik ecological legacy in transition. *Arctic Anthropology*, 34(1), 236–52.

Larsen, S.C. and Johnson, J.T. 2012. In between worlds: Place, experience, and research in indigenous geography. *Journal of Cultural Geography*, 29(1), 1–14.

Lefebvre, H. 2002 [1974]. *The Production of Space*. 19th Edition. Oxford: Blackwell.

Levinson, S.C. 2003. *Space in Language and Cognition: Explorations in Cognitive Diversity*. Cambridge: Cambridge University Press.

Lewis, D. 1972. *We the Navigators*. Canberra: Australian National University Press.

Lewis, D. 1976a. Observations on route finding and spatial orientation among the aboriginal peoples of the Western Desert region of central Australia. *Oceania*, 46, 249–82.

Lewis, D. 1976b. Route finding by desert aborigines in Australia. *Journal of Navigation*, 29, 21–38.

Livingstone, D. 2002. *Science, Space and Hermeneutics*. Heidelberg: University of Heidelberg.

Low, S.M. 2009. Towards an anthropological theory of space and place. *Semiotica*, 175(1–4), 21–37.

Lynch, K. 1960. *The Image of the City*. Cambridge, MA: MIT Press.

Massey, D. 1993. Power-geometry and a progressive sense of place, in *Mapping the Futures*, edited by J. Bird, B. Curtis, G. Robertson and L. Tickner. London: Routledge, 59–69.

Massey, D. 1994. *Space, Place and Gender*. Minneapolis: Polity Press.

Massey, D. 1999. Spacetime, 'science' and the relationship between physical geography and human geography. *Transactions of the Institute of British Geographers*, 24, 261–76.

Matless, D. 1997. The geographical self, the nature of the social and geoaesthetics: Work in social and cultural geography. *Progress in Human Geography*, 21(3), 393–405.

Mayhew, R.J. 2011. Geography's genealogies, in *SAGE Handbook of Geographical Knowledge*, edited by J.A. Agnew and D.N. Livingstone. London: SAGE Publications Ltd, 21–38.

Mazzullo, N. 2009. Sápmi: A symbolic re-appropriation of Lapland as Saamiland, in *Máttut – Máddagat: The Roots of Saami Ethnicities, Societies and Spaces/ Places*, edited by T. Äikäs. Oulu: Giellagas Institute, 174–85.

Merrifield, A. 2000. Henri Lefebvre: A socialist in space, in *Thinking Space*, edited by M. Crang and N. Thrift, London: Routledge, 167–82.

Metzo, K. 2009. Civil society and the debate over pipelines in Tunka National Park, Russia, in *Environmental Justice in the Former Soviet Union,* edited by J. Agyeman and Y. Ogneva-Himmelberger. Boston, MA: MIT Press, 119–40.

Miggelbrink, J. and Mazzullo, N.E. 2011. Winterweide und Holzlieferant: Indigene und andere Interessen im Konflikt über die Nutzung nordfinnischer Wälder [Winter pasture and timber provider: Indigenous and other interests in conflicts over utilization of northern Finnish forests]. *Geographische Rundschau*, 63(7–8), 36–42.

Mitchell, D. 1995. There is no such thing as culture: Towards a reconceptualization of the idea of culture in geography. *Transactions of the Institute of British Geographers*, New Series, 20, 102–16.

Morton, J. 2010. Why should governmentality matter for the study of pastoral development. *Nomadic Peoples*, 14(1), 6–30.

Nadasdy, P. 1999. The politics of TEK: Power and the 'integration' of knowledge. *Arctic Anthropology*, 36(1–2), 1–18.

Nash, C. 2000. Performativity in practice: Some recent work in cultural geography. *Progress in Human Geography*, 24(4), 653–64.

Nelson, R. 1969. *Hunters of the Northern Ice*. Chicago: University of Chicago Press.

Niezen, R. 2003. *The Origins of Indigenism: Human Rights and the Politics of Identity*. Berkeley, CA: University of California Press.

Paine, R. 1996. Saami reindeer pastoralism and the Norwegian state, 1960s–1990s. *Nomadic Peoples*, 38(2), 125–35.

Pearce, F. 2012. *The Land Grabbers: The New Fight over Who Owns the Earth.* Boston: Beacon Press.

Peet, R. 1985. Radical geography in the United States: A personal history. *Antipode*, 17(2–3), 1–7.

Peet, R. 1998. *Modern Geographical Thought*. Oxford: University Press.

Pennanen, J. and Näkkäläjärvi, K. 2003. *Siidastallan: From Lapp Communities to Modern Sámi Life.* Inari: Siida Sámi Museum.

Philo, C. 1995. Details, Verschiedenheiten und Geographie – Anmerkungen zur Geographie von Michel Foucault [Details, diversities and geography: Notes on the geography of Michel Foucault], in *Geographie und Gesellschaftstheorie: Referate im Rahmen des 'Anglo-Austrian Seminar on Geography and Social Theory' in Zell am Moos, Oberösterreich*, edited by W. Matznetter. Wien: Institut für Geographie und Regionalforschung der Universität Wien, 57–69.

Philo, C. 2000. Foucault's geography, in *Thinking Space,* edited by M. Crang and N. Thrift. London: Routledge, 205–38.

Pile, S. 2008. Where is the subject? Geographical imaginations and spatializing subjectivities. *Subjectivity*, 23, 206–18.

Porteus, S. 1937. *Primitive Intelligence and Environment*. New York: Macmillan.

Povorozniuk, O. 2011. Zabaikal'skie evenki: Sotsial'no-ekonomicheskie i kul'turnye transformatsii v XX–XXI vekakh [Evenki of Transbaikalia: Social-economic and cultural transformations in the twentieth and twenty-first century]. Moskva: Institut etnologii i antropologii Rossiiskoi Akademii Nauk.

Prager, L. (ed.). 2012. *Nomadismus in der Alten Welt* [Nomadism in the Old World]. Münster: LIT.

Price, M. and Lewis, M. 1993. The reinvention of cultural geography. *Annals of the Association of American Geographers*, 83(1), 1–17.

Raitio, K. 2008. 'You Can't Please Everyone': Conflict Management Practices, Frames and Institutions in Finnish State Forests. Joensuu: University of Joensuu.

Reichert, D. 1996. Räumliches Denken als Ordnen der Dinge [Spatial thinking as an arrangement of things], in *Räumliches Denken*, edited by D. Reichert. Zürich: Verlag der Fachvereine, 15–46.

Rethmann, P. 2004. Domesticity, change and gender in northern Kamchatka. *Anthropology of East Europe Review*, 28(2), 85–101 [Online]. Available at: http://scholarworks.iu.edu/journals/index.php/aeer/article/view/932/1040 [accessed: 18 November 2012].

Reuning, H. 1988. Testing Bushmen in the central Kalahari, in *Human Abilities in Cultural Context*, edited by S. Irvine and J. Berry. Cambridge: Cambridge University Press, 453–86.

Rose, G. 1993. Some notes towards thinking about the spaces of the future, in *Mapping the Futures*, edited by J. Bird, B. Curtis, G. Robertson and L. Tickner. London: Routledge, 70–83.

Rose-Redwood, R.S. 2006. Governmentality, geography and the geo-coded world. *Progress in Human Geography*, 30(4), 469–86.

Ruotsala, H. 2008. Does sense of place still exist? *Journal of Ethnology and Folkloristics* 1(2), 43–54 [Online]. Available at http://www.jef.ee/index.php/journal/article/view/36/pdf_35 [accessed: 18 November 2012].

Ruotsala, H. 2011. Ancestors' wisdom or desktop reindeer management? The role of traditional ecological knowledge in contemporary reindeer herding, in *Thinking Through the Environment: Approaches to History*, edited by Timo Myllyntaus. Cambridge: White Horse Press, 159–78.

Salzman, P.C. (ed.). 1980. *When Nomads Settle: Processes of Sedentarization as Adaptation and Response*. New York: Praeger.

Sasaki, S. 2010. Establishment of large-scale reindeer herding in the European and West Siberian tundra, in *Good to Eat, Good to Live with: Nomads and Animals in Northern Eurasia and Africa*, edited by F. Stammler and H. Takakura. Sendai: Center for Northeast Asian Studies, Tohoku University, 77–99.

Sayer, A. 2000. Critical and uncritical cultural turns, in *Cultural Turns/ Geographical Turns: Perspectives on Cultural Geography*, edited by I. Cook, D. Crouch, S. Naylor and J.R. Ryan. Harlow: Prentice Hall, 166–81.

Schelhaas, B. and Hönsch, I. 2001. History of German geography: Worldwide reputation and strategies of nationalisation and institutionalisation, in *Geography: Discipline, Profession and Subject since 1870 (An International Survey)*, edited by G.S. Dunbar. Dordrecht: Kluwer, 9–44.

Schmajuk, N.A. and Thieme, A.D. 1992. Purposive behavior and cognitive mapping: A neural network model. *Biological Cybernetics*, 67(2), 165–74.

Schmid, C. 2005. *Stadt, Raum und Gesellschaft: Henri Lefebvre und die Theorie der Produktion des Raums* [City, space and society: Henri Lefebvre and the theory of production of space]. Stuttgart: Franz Steiner.

Scholz, F. (ed.). 1991. *Nomaden – Mobile Tierhaltung: Zur gegenwärtigen Lage von Nomaden und zu den Problemen und Chancen mobiler Tierhaltung* [Nomads – mobile pastoralism: On the current situation of nomads and the problems and prospects of mobile pastoralism]. Berlin: Reimer.

Scholz, F. 1995. *Nomadismus: Theorie und Wandel einer sozio-ökologischen Kulturweise* [Nomadism: Theory and change of a social-ecological mode of culture]. Stuttgart: Franz Steiner.

Scholz, F. 1999. Ist Nomadismus tot? Mobile Tierhaltung als zeitgemäße Nutzungsform der kargen Weiden des Altweltlichen Trockengürtels [Is nomadism dead? Mobile pastoralism as a contemporary form of utilizing meagre pastures in the dry belt of the Old World]. *Geographische Rundschau*, 51(5), 248–55.

Schultz, H.-D. 2008. Vorletzte Antworten auf letzte Fragen zur Klassischen Geographie: Eine Reaktion auf Ulrich Eisels Kritik [Penultimate answers to ultimate questions of classical geography: A response to Ulrich Eisel's critique], in *Klassische Geographie: Geschlossenes Paradigma oder variabler Denkstil?*, by U. Eisel and H.-D. Schultz. Berlin: Humboldt-Universität, 39–276.

Scott, C. 2004. 'Our Feet are on the Land, but our Hands are in the Sea': Knowing and Caring for Marine Territory at Erub, Torres Strait, in *Woven Histories, Dancing Lives: Torres Strait Islander Identity, Culture and History*, edited by Richard Davis, Canberra: Aboriginal Studies Press, 259–70.

Scott, C. and Mulrennan, M. 2010. Reconfiguring Mare Nullius: Indigenous Sea rights and the divergence of domestic and international norms, in *Indigenous Peoples and Autonomy: Insights for a Global Age*, edited by M. Blaser, R. de Costa and W. Coleman, Vancouver: University of British Columbia Press, 148-176.

Simonsen, K. 1996. What kind of space in what kind of social theory?, in: *Progress in Human Geography* 20, H. 4, 494–512.

Slezkine, Y. 1994. *Arctic Mirrors: Russia and the Small Peoples of the North.* Ithaca, NY: Cornell University Press.

Smith, N. 1984. *Uneven Development: Nature, Capital and the Production of Space*. Oxford: Blackwell.

Smith, N. and Katz, C.1993. Grounding metaphor: Towards a spatialized politics, in *Place and the Politics of Identity*, edited by M. Keith and S. Pile. London, New York: Routledge, 67–83.

Sneath, D. 2004. Proprietary regimes and sociotechnical systems: Rights over land in Mongolia's 'age of the market', in *Property in Question: Value Transformation in the Global Economy*, edited by K. Verdery and C. Humphrey. Oxford: Berg, 161–82.

Sneath, D. and Kaplonski, C. (eds). 2010. *The History of Mongolia*. 3 Volumes. Folkestone: Global Oriental.

Soja, E.W. (ed.). 1989. *Postmodern Geographies: The Reassertion of Space in Critical Social Theory*. London: Verso.

Sokolovski, S.V. 2007. Indigeneity construction in the Russian Census 2002. *Sibirica*, 6(1), 59–94.

Stammler, F. 2005a. *Reindeer Nomads Meet the Market: Culture, Property and Globalisation at the 'End of the Land'*. Münster: LIT.

Stammler, F. 2005b. The obshchina movement in Yamal: Defending territories to build identities?, in *Rebuilding Identities: Pathways to Reform in Post-Soviet Siberia*, edited by E. Kasten. Berlin: Reimer, 109–34.

Stammler, F. 2009. Mobile phone revolution in the tundra? Technological change among Russian reindeer nomads. *Folklore*, 41, 47–78 [Online]. Available at http://www.folklore.ee/folklore/vol41/stammler.pdf [accessed: 18 November 2012].

Stammler, F. and Eilmsteiner-Saxinger, G. (eds). 2010. *Biography, Shift-Labour and Socialisation in a Northern Industrial City – The Far North: Particularities of Labour and Human Socialisation* (Proceedings of the International Conference in Novy Urengoy, Russia, 4–6 December 2008). Rovaniemi: Arctic Centre, University of Lapland [Online]. Available at: http://arcticcentre.ulapland.fi/docs/NURbook_2ed_100421_final.pdf [accessed: 18 November 2012].

Stammler, F. and Peskov, V. 2008. Building a 'culture of dialogue' among stakeholders in north-west Russian oil extraction. *Europe–Asia Studies*, 60(5), 831–49.

Stanek, L. 2011. *Henri Lefebvre on Space: Architecture, Urban Research, and the Production of Theory*. Minneapolis: University of Minnesota Press.

Thompson, N. 2008. *Settlers on the Edge. Identity and Modernization on Russia's Arctic Frontier*. Vancouver: UBC Press.

Thornton, T. 2008. *Being and Place among the Tlingit*. Seattle: Washington University Press.

Ulturgasheva, O. 2012. *Narrating the Future in Siberia: Childhood, Adolescence and Autobiography among Young Eveny*. New York: Berghahn.

Van den Scott, L.-J. 2009. Cancelled, aborted, late, mechanical: The vagaries of air travel in Arviat, Nunavut, Canada, in *The Cultures of Alternative Mobilities: Routes Less Travelled*, edited by P. Vannini. Farnham: Ashgate, 211–26.

Ventsel, A. 2005. *Reindeer, Rodina and Reciprocity: Kinship and Property Relations in a Siberian Village*. Münster: LIT.

Vermeulen, H. 2011. Linguistik und Völkerkunde: Der Beitrag der historisch-vergleichenden Linguistik von G.W. Leibniz zur Entstehung der Völkerkunde im 18. Jahrhundert [Linguistics and ethnology: The contribution of G.W. Leibniz's comparative historical linguistics to the emergence of ethnology in the eighteenth century]. *Max Planck Institute for Social Anthropology Working Papers*, 133 [Online]. Available at: http://www.eth.mpg.de/pubs/wps/pdf/mpi-eth-working-paper-0133.pdf [accessed: 1 November 2012].

Vitebsky, P. 2005. *Reindeer People: Living with Animals and Spirits in Siberia*. London: HarperCollins.

Vladimirova, V. 2006. *Just Labor: Labor Ethic in a Post-Soviet Reindeer Herding Community*. Uppsala: Uppsala Universitetet.

Weichhart, P. 1993. Vom 'Räumeln' in der Geographie und anderen Disziplinen: Einige Thesen zum Raumaspekt sozialer Phänomene [On the 'spatializing' in geography and other disciplines: Some theses on spatial aspects of social phenomena], in *Die aufgeräumte Welt: Raumbilder und Raumkonzepte im Zeitalter globaler Marktwirtschaft*, edited by J. Mayer. Rehburg-Loccum: Evangelische Akademie Loccum, 225–42.

Weichhart, P. 1999. Die Räume zwischen den Welten und die Welt der Räume [The space between worlds and the world of spaces], in *Handlungszentrierte Sozialgeographie: Benno Werlens Entwurf in kritischer Diskussion*, edited by P. Meusburger. Stuttgart: Franz Steiner, 67–94.

Weiers, M. (ed.). 1986. *Die Mongolen: Beiträge zu ihrer Geschichte und Kultur* [The Mongols: Contributions on their history and culture]. Darmstadt: Wissenschaftliche Buchgesellschaft.

Wenzel, G.W. 1999. Traditional ecological knowledge and Inuit: Reflections on TEK research and ethics. *Arctic*, 52(2), 113–24.

Widlok, T. 1997. Orientation in the wild: The shared cognition of Hai||om Bushpeople. *Journal of the Royal Anthropological Institute*, New Series, 3(2), 317–32.

Wießner, R. 1978. Verhaltensorientierte Geographie: Die angelsächsische behavioral geography und ihre sozialgeographischen Ansätze [Behaviour-oriented geography: Anglo-Saxon behavioural geography and its social-geography approaches]. *Geographische Rundschau* 30 (11), 420–26.

Wiget, A. and Balalaeva, O. 2011. *Khanty: People of the Taiga – Surviving the 20th Century*. Fairbanks: University of Alaska Press.

Withers, C.W.J. 2011. Geography's narratives and intellectual history, in *SAGE Handbook of Geographical Knowledge*, edited by J.A. Agnew and D.N. Livingstone. London: SAGE Publications Ltd, 39–50.

Woodward, K., Jones III, J.P. and Marston, S.A. 2010. Of eagles and flies: Orientations towards the site. *Area*, 42(3), 271–80.

Yiftachel, O. 2006. *Ethnocracy: Land and Identity Politics in Israel/Palestine*. Philadelphia, PA: University of Pennsylvania Press.

Zenker, O. 2011. Autochthony, ethnicity, indigeneity and nationalism: Time-honouring and state-oriented modes of rooting individual-territory-group triads in a globalizing world. *Critique of Anthropology*, 31(1), 63–81.

Ziker, J. 2002. *Peoples of the Tundra: Northern Siberians in the Post-Communist Transition*. Prospect Heights, IL: Waveland Press.

Chapter 2
A Place Off the Map: The Case for a Non-Map-based Place Title

Denis Wood

Who needs to know my location? Strangers do, outsiders, others. Location is always about others. It's about me to them. It's about them to me. It's about them to each other.

I'm standing in the front yard raking leaves from the little strip of grass separating the sidewalk from the street. The leaves are a mix of pin oak, crepe myrtle and maple. Looking up, I notice a car coming slowly down the street. It may not make much sense to talk about the body language of a car, but there's something about the way this car's moving that says it's lost.

I can see a woman behind the wheel. She slows to a stop in front of me and powers down the window.

'Excuse me', she says.

I walk out into the street and lean down toward her window. There's a young girl in the seat beside her.

'Can you help me?' she asks. 'I have no idea where I am'.

I want to say, 'Sure, sure I can help you. You're *here*!' but I know she doesn't really want to know where she is. She wants to know where she is *vis-à-vis somewhere else*. She wants to know how to get out of here. So instead I ask her, 'Where are you trying to go?' and I tell her how to get there.

Where *I* am, here, my *place*, it's not the same thing as my location. My location is my place *vis-à-vis someplace else*. If my *place* is local, is here, has crepe myrtles and pin oaks and maples in it, my *location* is abstract and is caught up in a universal grid. Most of us – well, I don't know that. I was going to say 'most of us live easily with this duality, can shift from the one to the other as the communication situation demands', but I don't know how many of us can easily jump off and on the map, or how meaningfully, or at what cost any do.

On the map ... It's not somewhere you need to be just because you've got a place in the world.

Two Places

Uncle Herman, paterfamilias in Ludwig Bemelmans's first book, *Hansi*, is a forest ranger in the Tyrolese Alps above Innsbruck. The house he lives in was built by his great-grandfather:

> His grandfather had lived in it – so had his father. He himself was born here and so were his children. The house was two hundred years old, and had always carried the name of his family. It was carved into the strong beam in front of the house where the balcony rested. They could never think of any other place in the land as their home.
>
> Mountain houses are fine and simple because they have grown from the rock on which they stand, from the forests that are around them, and from the work of men who looked at mountains all their lives and to whom every tree and flower said, 'See how lovely we are in delicate colors and strong clear patterns'. (Bemelmans 1934: 24)

The house has actually *grown* from the rock. The family has *always* lived in it. The family, the house are exemplars of the rooted. Even the furniture is of the place, has grown from it: 'Each piece was made by hand – no two were alike. Someone had sat down in a room, looked out the window, and said, "I'll build a bed for this room, or a chair." It fitted and stood in its place from then on' (Bemelmans 1934: 24). *In its place*: its situation is as indisputable as the fact that Uncle Herman lives *here*, lives *some*where.

Are there really people today who live where their great-grandfathers did? Sure. Four generations is not that hard to pull off, especially if we loosen the strictures some, if not in the same *house*, then say in the same parish, town, or county. There must be lots of such people. (Though it *would* be interesting to know how many, and how long any have lived anywhere.)

Contrast Uncle Herman with Ryan Bingham, the George Clooney character in the film, *Up in the Air*. Bingham makes his living traveling around the United States to fire people whose employers lack the guts to do it themselves. As a sideline, he delivers motivational speeches that extol a life free of both things and entangling relationships. Bingham luxuriates in the anonymity of his perpetual travel. He *loves* airport lobbies, indistinguishable hotel rooms, his suitcase – he's miserable when he's temporarily grounded – and he holds as an overriding ambition the accumulation of ten million frequent-flyer miles. Ryan Bingham is *nowhere man* incarnate. Despite this he too has a *place*, even if it's spread here and there all over the country: when at the film's end Bingham once again stands in front of a departures and arrivals board in the middle of a busy airport concourse, the satisfied expression that crosses his face makes it perfectly plain that he's home.

Is Ryan Bingham a fiction? Hard to say. Pilots, stewardesses, long-haul truck drivers, ships captains, migrant farm workers, salespeople may be other examples of those with nowhere places, '*may* be', I say because any number of these may also have places to which they return again and again, *some*where places, places where they do feel rooted. More like Bingham, others may not. I mean, I've known academics *prone to going on about the virtues of rootedness*, who were so busy juggling their many appointments, so busy jetting from one conference to another, that I can't believe that like Bingham they weren't more at home on the road or in the air, settling into their seat on an airplane, checking into a hotel. And the next day into another.[1]

Now, we can assign a *location* to Uncle Herman's place in the world but we can't do that for Bingham's. At best we could assign Bingham's place to a string of locations, though in addition to the ones he's visited already they'd have to include those he has yet to, for unlike Uncle Herman's place which, whatever else it is, is also a *site*, Bingham's place is better thought about as something like a *niche* which, given that a niche is an ecological, not a geographical concept, cannot be posted to a map.

Place Is Like a Niche

'By niche', John Tyler Bonner says (1969: 61, emphasis mine), is meant 'the *place* in nature of the organism. The important emphasis, and in fact the value of the concept of the niche, is that it pinpoints the function, the activity, of the organism within its environmental community. It designates what the animal or plant does rather than what it looks like' (where in 'animal' I want you to hear 'Ryan Bingham').[2] In *Animal Ecology*, the book that established the paradigm of the niche for modern times, Charles Elton wrote that when an ecologist sees a badger 'he should include in his thoughts some definite idea of the animal's place in the community to which it belongs, just as if he had said "there goes the vicar",' (Elton quoted in Hutchinson 1978: 157) where by 'place' should be understood 'the many ranges of conditions and resource qualities within which the organism or species persists, often conceived as a multidimensional space' (Ricklefs 1990: 817). Again, this isn't a physical, it's an abstract space in which coordinates are defined by the values of continuously varying resource attributes, typically things like temperature, insolation, humidity, soil particle size, branch density, nutritional value and the like (Ricklefs 1990: 728–47); though in Bingham's case we'd have to add the availability and quality of airline club lounges (such American Airlines Admirals Clubs), first class accommodations, express lanes, complimentary beverages, turn down services and so on.

---

1 Notoriously, in a review of Anne Buttimer and David Seamon's *The Human Experience of Space and Place*, I described Buttimer as being one of these peripatetic professors. Irritated by her condescension towards the Worcester residents she imagined 'scarcely ever thought about place at all', I wrote: 'And instead of imagining that this might be because of her bizarre theories or unwarranted assumptions – or because they [the residents] have, in their strong sense of place, no need to belabour endlessly what she in her ceaseless jetting among Ireland, Sweden, and the United States has not – she concludes that they "had become much better adapted to placelessness and individualism than I"' (Wood 1982: 503–6). The arrogance of her words still infuriates me.

2 Of course Bingham constructs his niche – and how to say this – more *actively* than a plant or most other animals can (or do); or perhaps one could say that his agency was greater. Which is not to deny the niche-creating agency of every living thing.

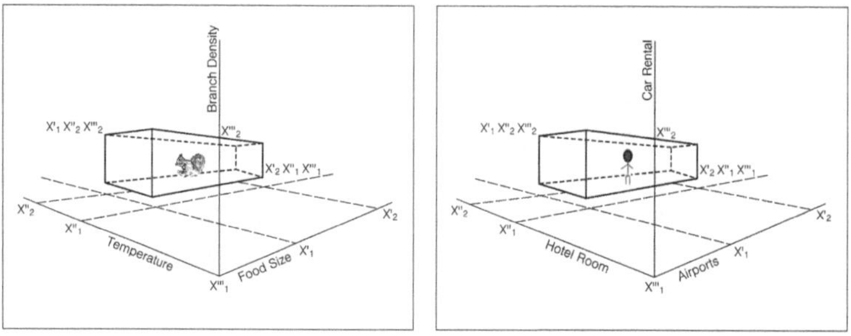

Two three-dimensional orthogonal fundamental niches, for a squirrel (left) and Ryan Bingham (right). The critical dimensions for the squirrel might include food size, temperature tolerance, and some measure of the density of branches; for Bingham they might include the accessibility of airports, hotel rooms, and rental cars (after Hutchinson 1978: 159).

Figure 2.1 The niche

For these niches to be occupied by actual organisms, however, the niches – *this* temperature range, *that* degree of salinity; *that* level of service, *this* degree of complaisance – have to be afforded, expressed, exhibited by or in *habitats*. One speaks of stream habitat, forest habitat, desert habitat, but plainly, there is also business-class airline-travel habitat, truck-stop habitat, labour-camp habitat. Habitats do exist physically and so they can be posted to maps, though whether the niche afforded by a given habitat is occupied is a question only inspection can determine. And whether those inhabiting the niche include a given *individual* – Ryan Bingham – is another question again.

I am *not* claiming that Uncle Herman doesn't have a niche. Unquestionably he does – he's a forest ranger, a householder, a father, an uncle – and he occupies the niche afforded by the mountain forest habitat above Innsbruck in the Tyrolese Alps. I insist, however, that his niche is *uniquely* afforded by *that* individual habitat, the one in the Tyrol above Innsbruck. This, it seems to me, is precisely the burden of the claim that the house *grew* from the rock it stands on, that the furniture is a response to the view from a window: that the house – and by extension Uncle Herman – is *there* and could only *be* there. Because it is exclusively afforded by a unique habitat, Uncle Herman's niche can't be distinguished from its habitat and so his niche effectively *has* a location.

Bingham's doesn't. Bingham's niche is expressed by habitats all over the country, hell, all over the world, with more being built daily (all it takes is a decent hotel, concierge service, a fruit basket in the room). The best we can speak of in Bingham's case is *potential range* which, with respect to habitat, is the entire geographical area containing suitable habitats. Range comes to something like a quantum wave function, indicating that the organism in question may be found somewhere within it but neither where nor how commonly. Where Bingham is at any given moment is lost in the cloud of probabilities vouchsafed by his place, that worldwide constellation of airports and hotels.

This is not to say that with a GPS we couldn't locate Bingham or that Bingham and the others lack an address. Every one of them – most of them anyway – has what we call a 'permanent address'. I mean, it's hard to get a (legal) driver's license without a permanent address; the taxman requires an address and because of this employers do too; you need an address for a passport, visa, work permit; these days you even need one to buy an airline ticket. But permanent addresses are often convenient fictions – a kind of *résidence actuelle de guerre* – where convenience, *bureaucratic* convenience, is the name of the game. To pretend our Binghams live at their permanent addresses, that they regard them as *home* – as their place – is to participate in a fraud only a bureaucrat could stomach.

Though Uncle Herman has little need of a permanent, or for that matter, any kind of address (like location, addresses too are for strangers), he very much lives at his, an address permanent in ways a Bingham could scarcely imagine (and certainly not appreciate). For an Uncle Herman, location, permanent address, niche, habitat, range, all are just different ways of naming the same thing: *Uncle Herman's place in the world.* For Bingham these are all very different. His location could be in the air somewhere between Seattle and Denver, his permanent address an apartment in Omaha, his niche that of privileged traveller, his habitat those of airports and hotels, his range the U.S., the world. None of these really catches his *place* which is not just the sum of his habitats but also *his characteristic moving among them*.

Two places, then: one readily posted to a map, one not; two milieus, one here, one all over … all over the *place*, we'd say in English, if that weren't piling ambiguity on top of ambiguity. Two places: one consisting at its core in private property (Uncle Herman's house and land), the other essentially in rented property (Bingham's hotel rooms, airplane seats); one in which the occupant has rights (all the rights of an owner), the other in which he doesn't, or damn few (and these printed in type too small to read on the back of a ticket). Now, Uncle Herman and Bingham may both be fictions, but in fact millions and millions of people live the way they do, and everyone is more or less rooted, more or less in flight; and the bureaucratic structure of our system of states very much privileges the people of the root over the people of the wing. Bluntly, people in motion are a threat. At the very least they lack a useful address, which is to say, they're hard to post to a map. This renders them hard to keep track of (so who knows what they're up to?) and lets them slip through the net in too many ways.

Maps Are Machines for Establishing Locations

The list I enumerated of entities requiring addresses – license bureaus, the taxman, immigration control, airlines – tells us something about addresses, in fact, about locations generally, namely that they're not primarily for the convenience of strangers, but for the convenience of that omnipresent stranger, the state. Let me say yet again what I say when I say 'location'. I'm not referring to the phenomenological experience of being somewhere. I'm always where I am, *here* in fact, wherever I am. Establishing my *location* means knowing where I am *in relation to somewhere else*. You're looking for your friend in a crowd. You can't see him. You call him on your cell phone: 'Where are you?' The two of you establish a framework – 'See the blue-striped awning?' – by setting a third point that can be used as a guide – 'I'm right across from it'. Now it's you, your friend, and the awning. The awning not only works as a reference, it transforms what were two *points* into a *space*. You, your friend, the awning, *and* that building with the steeple, and the tall skinny tree, and the balloon man: you add enough of these and you can make a map of the world. 'So how do I get to your house?' a new acquaintance asks. You reel off the sequence of moves that will get him to you, or you refer to that general system of collocations we call the map and just give him an address. Type 'location' into Google, and Google Maps comes back at the head of the list. MapQuest comes second.

Maps *can* be thought about as systems of collocations like these, and maps love this sort of alibi, you know, that they're nothing but simple things out to make life easier for us. Interviewing me about maps, Ira Glass, the host of WBEZ Chicago's *This American Life*, asked about the vast number of maps I claimed had been made in the twentieth century: 'What are those maps', he wondered, 'and what proportion of them do you think are the maps that most of us, civilians, usually use which are just road maps to get us from one place to another?'

I cut him off: 'No, but you see, I think you've missed all the maps right off the bat as soon as you go to the road map, because you've forgotten …' and we followed this with the beginnings of a catalogue – the weather map, the maps in newspapers – which would have ended where mapmaking began, with the interest of proto-states in controlling the ownership of property. As we know it today, mapmaking is a kludgy technology, cobbled together in the fourteenth and fifteenth centuries (maybe the twelfth century in China) from a grab bag of previously independent discourse functions. One of these was certainly this very large-scale, graphic, property control function, documentable to 2300 BCE in Babylon, to the eighth century CE in Japan, and to the twelfth through the fifteenth centuries CE in England. Wholly unrelated to this was a rarer, very small-scale, cosmographic speculation function that can be documented from equally disparate times and places, from the well-known 'Babylonian World Map' of c. 600 BCE, for instance, through medieval European *mappaemundi*, to the Buddhological world maps made as early as the fourteenth century CE in Japan. There was also a relatively small-scale, coastal-navigation function that emerged during the late medieval period that seems equally unrelated to either the property control or cosmographic speculation functions. (In China there may also have been a military planning function.) In the fourteenth and fifteenth centuries these very different discourse functions, together with others entirely novel, began to be understood as no more than different faces of a generalizable *locational discourse function* – the map – into which, over succeeding centuries, more and more of life has gradually been drawn.[3]

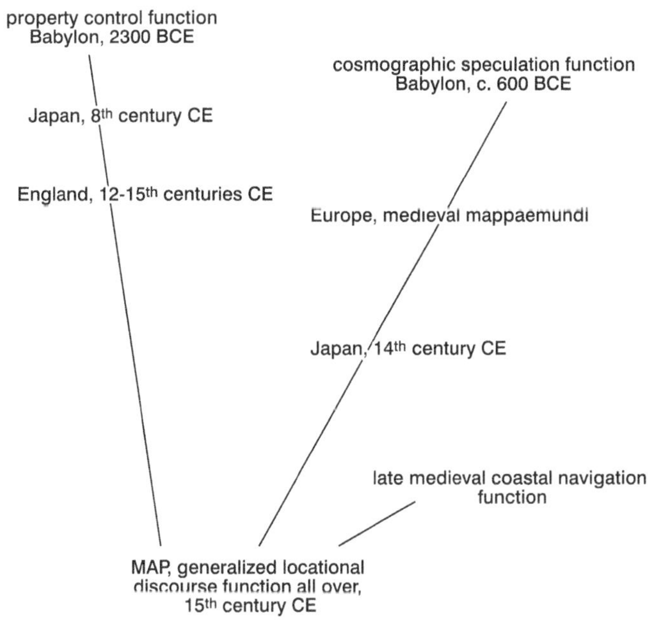

The map, as the generalized locational discourse function we recognize today, emerged in the fifteenth century, in response to the needs of the early modern state, out of more specific discourse functions including property control, cosmographic speculation, and coastal navigation.

Figure 2.2 The origin of the map

3 Note that I don't allude to a fourth, even more marginal function, which we might describe as genuinely *geographic* – or, perhaps more precisely, Ptolemaic – despite the fact that although it dies just as mapmaking is beginning to become widespread, its ghost haunts the very idea of the map for centuries.

It's probably not necessary to have said more than 1) prior to the fourteenth and fifteenth centuries few people used maps at all, and none used them for much; and 2) the revisioning of what had been separate strands as but fibers of a common *map thread*, occurred at the same time that polities around the world began to understand themselves as modern, or proto-modern, states. Because these new states construed themselves as territorial polities – in contradistinction, say, to those feudal societies organized around bonds of reciprocal obligations out of which so many young states emerged – states discovered a huge interest in location and so in mapmaking (Wood 2010: 27–35). Indeed it may not be too much to say that modern states consist of little more than great tabulations of locations, increasingly in map form, over which states exercise their various authorities: maps of the locations of the territory over which they are sovereign, and therefore maps of the locations of the borders which bound their territories; maps of the locations of their constituent elements (territories, provinces, states, counties, parishes and the like, each in turn making maps like crazy); maps of the locations of their resources and properties (which is to say *all* resources and properties over which they exercise eminent domain); maps of the locations of their citizens (to deliver services, question, tax, conscript); as well as maps of the locations of all the things *outside* their boundaries that concern them, which in the case of large modern states is almost everything in the world.

To say nothing of the moon. And Mars. And the rest of the solar system.

Everything that is, *that has a location,* for the logic of the map is a propositional one constructed out of what John Fels and I call 'postings', fundamental cartographic propositions that *this is there* (Wood and Fels 2008: 26–33). Each of these postings encapsulates a powerful existence claim – *this is* – that gains enormous power by being *posted* (that is, from the indexicality vouchsafed by the sign plane of the map). The power gained by the posting of these existence claims arises from the fact that every instance of map use constitutes an implicit act of validation. This validation – all but automatic – is structured by antecedent validations performed in situations ranging from map-learning exercises in school, through successful uses of maps in way finding, to the sight of Colin Powell pointing out on a map of Iraq the locations of weapons of mass destruction.[4] The claim *this is there* is powerful precisely because it implies the performance of an existence test: *that you can go there and check it out*. Having done this in the past, you know the outcome: *it is there*. Besides, who would fake such a challenge? The assent thus given to the postings spreads to the territory that the postings collaboratively construct, and this endows the

4 The whole process is similar to what Steven Shapin and Simon Schaffer (1985) write about as the production of facts by witnessing and reporting in their *Leviathan and the Air-Pump*.

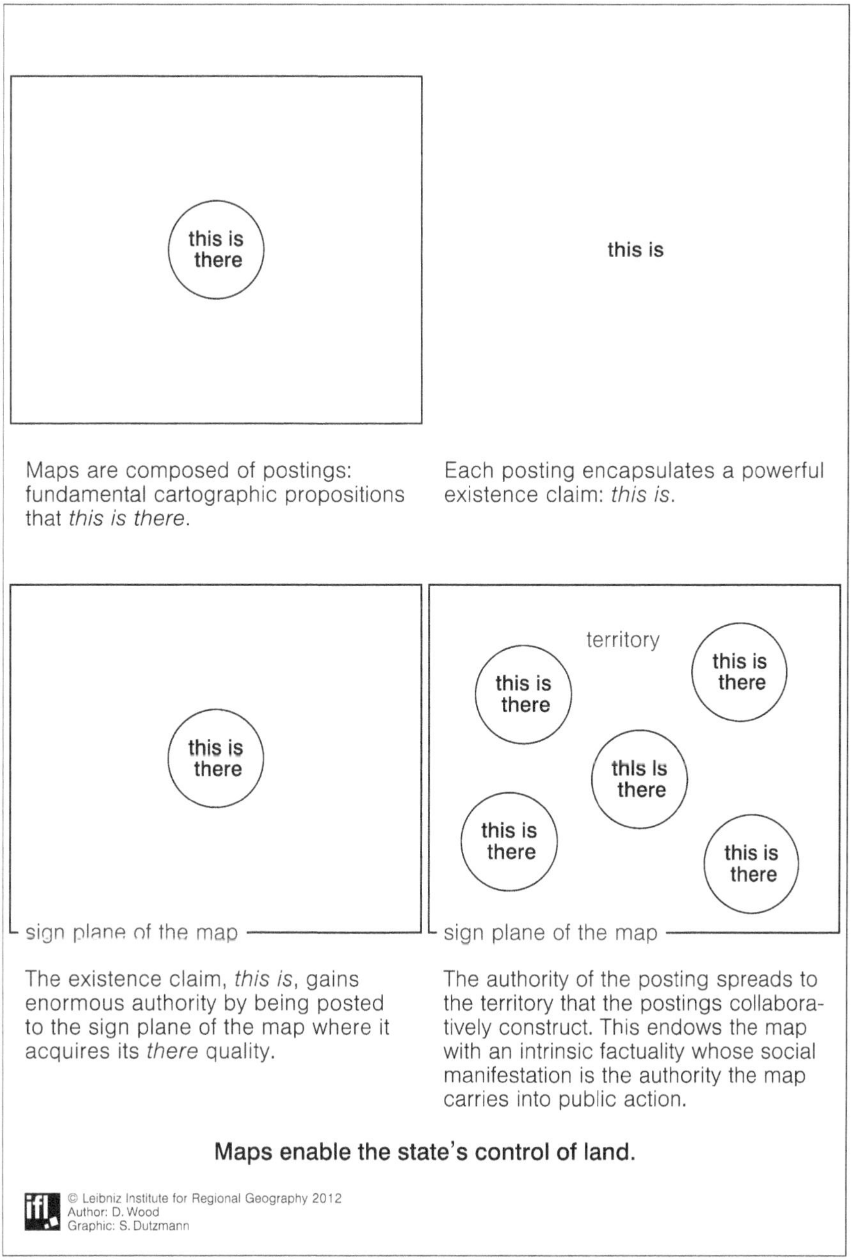

Figure 2.3 This is there

map as a whole with an intrinsic factuality whose social manifestation is the authority the map carries into public action.[5]

Can it be doubted that this locative authority is the reason that the earliest and most consistent use of what became maps – across cultures and throughout history – is *the control of land, the registration of real property*? I think not. Nor, I think, can it be doubted that it was this locative authority that gave the map so heightened a role in the rise of the early modern state. In their history of cadastral mapping Roger Kain and Elizabeth Baigent (1992: dust jacket flap) put it this way: 'Cadastral maps played an important role in the rise of modern Europe' – and I might add modern Asia, the modern Americas, and Australia – 'as tools for the consolidation and extension of land-based national power', where by 'extension' we need to hear among many other things ... *colonial settlement*. Kain and Baigent (1992: 265) go on to say:

> In the early years of European settlement in the New World in the seventeenth century, whether in the Liesbeeck River Valley east of Cape Town in South Africa, or on the Atlantic seaboard of North America, land surveying and the production of cadastral maps became established as a concomitant of colonial settlement. Land availability, if not the only lure of migrations from Europe, was a most important influence in the individual decision to migrate. As Sarah Hughes comments in the context of Virginia: 'Immigrant colonists gazing at a wilderness envisaged its taming and imagined new markers bounding the edges of their own fields and meadows. The men who could measure the metes and bounds of those fields held the key to transforming a worthless, uncultivated territory into individual farms [...]'

'Individual farms', 'the metes and bounds of those fields', 'markers bounding their own fields and meadows', may not be words Uncle Herman ever uttered, but the idea of place that these words encapsulate is one he'd understand: a place that was simultaneously a location, a location that could be, when necessary – and almost invariably was – posted to a map. It's no surprise that non-Han peoples who declared their loyalty to the Chinese state were said 'to enter the map' (Csete 2006: 235–37, 240). There were, of course, differences in the processes of settlement and their relationships to cadastral mapping between Capetown and Jamestown, between Jamestown and Plymouth, but common to them all was the reality that the map was the machine that established the locations, and absolutely *nothing* like a sketch drawn 'to communicate a sense of place, some sense of *here* in relation to *there*' (Wilford 2000: 6).[6]

5 The use of the map by Colin Powell to advance the Bush war claims is a brilliant example of how the map's authority can be/is exploited in public action. Bush may have intended to go to war whatever the case, but the authority of the Powell map greased the political skids.

6 This is another favoured alibi of maps, that they're all about place.

'To communicate a sense of place': this is reaching for an idea of the map as a poetics of place, and perhaps a poetics of place is not wholly beyond the map's reach; but it had nothing to do with the maps made of the parts of the world soon to be seized, and then transformed, by European colonizers. The maps the migrants saw said, 'Here there is land', but little about place, that is, little about its conditions, about the winter cold or summertime heat, about the unfamiliar fauna, about the numbers and sizes of the mosquitoes, about the strange, often bizarre vegetation, and *extraordinarily little* – or as little as could be gotten away with – about the humans already occupying the land.

I mean, what sort of *place* are you describing when you omit the people living there?

But maps are good at that. You know, it's lat/long here and lat/long there, so many hectares, here a river, there a swamp; and for this or that consideration, the patenting of so much land. And, whoops, *you're who? You* live *here? Since when?*

Since always? But this is my land ...

Immigrants, settlers, colonists, they weren't much good at recognizing aboriginal title. After all, it was invariably customary in form and so had never been patented, which is a way of saying that it was place-, not location-based. Immigrants were particularly bad when it came to mobile swidden cultivators, to indigenous occupants who cultivated less than they foraged, to hunters, to herders. Their places were too much like Ryan Bingham's, and the immigrants knew about nothing but places like Uncle Herman's.

These are not coincidences, the behaviour of the colonists and the characteristic ineptitude of maps, for if maps were great at establishing location and pathetic when it came to expressing a sense of place, it was precisely this pair of complementary 'virtues' that made the map so invaluable in laying the grounds for migration. A sense of *place* would only have ... *gotten in the way*, could only have *deterred* people from imagining a life of their own in a place already richly inhabited by others. Only when maps are understood in this way does Bernard Nietschmann's (1995: 34–7), 'More Indigenous territory has been claimed by maps than by guns', make any sense, though at the very same time it renders wholly moot his assertion that, 'and more Indigenous territory can be reclaimed and defended by maps than by guns'.[7]

In the first place, as I've pointed out elsewhere, maps themselves have no power at all (Wood 2010: 1–7, 117–18). They're rather used to *wield* power: power flows *through* the map. Power is a measure of work, and work is the application of a force through a distance. The work of maps is to apply social forces to people to bring into being a particular socialized space. The forces in question? *Ultimately*,

7 And Nietschmann (1995: 37) went on: 'Whereas maps like guns must be accurate, they have the additional advantages that they are inexpensive, don't require a permit, can be openly carried and used, internationally neutralize the invader's one-sided legalistic claims, and can be duplicated and transmitted electronically which defies all borders, all pretexts, and all occupations'.

I've insisted, they're those of the courts, the police, and the military. The reason maps are so often turned to is because of their ability to replace, *to reduce the necessity for*, the application of armed force. For armed force maps substitute the force of the authority of the map, *but the map's authority cannot be separated from that of the state that backs it up*. Put simply, the authority of the map is only as great as the authority of the state that guarantees it and only in the rarest cases is a state about to guarantee maps securing land claims against it.

... But Not of Place

I think that's an unanswerable objection – international approbation and goodwill come to nothing in the face of an intransigent state determined to defend its authority (*vide* Israel) – but more critical to my mind is the misfit between the map and place when the place is Bingham-like. At the moment I'm thinking about Travellers. The classic examples are the Gypsies, the Roma, the Romani, but there are all kinds of Travellers. Government statistics would have you believe that most have been 'settled' but too many of them too often fail the signal test of an address, of a location you can point to on a map: when you go there to check it out ... they're *not* there.

We're not talking about a couple of people either. Romani live all over the world. There may be none in Antarctica, but every other continent has its share, though what that share is is unknown, even poorly estimated; partly because it's unclear who should be counted (are the Yeniche Romani or not?), partly because few of them are eager to be counted in the first place. Estimates laughably range from two to 14 million for Europe and Asia Minor where perhaps most of them live in sizeable numbers in almost every country: the Italian Usari; the Roma, the Sinti of Germany, Austria and eastern Europe; the Yeniche, Jenische, the 'White Gypsies' of Germany, Switzerland and France; the Reisende, or Indigenous Norwegian Travellers, who may or may not be confused with the Tater, or Norwegian Travellers; the Finnish Kale, maybe a quarter of them living in Sweden; the half-dozen different Scottish Traveller groups; the various Tinkers or Irish Travellers; the Welsh Kale; the Romanichal of England and the Scottish Borders; and others still in Portugal, Spain, Greece, Hungary, Romania, Bulgaria, Russia, Turkey.

Besides these are Occupational Travellers: circus, carnival and fairground workers and their families; bargee and other waterway workers; migrant farm workers and other seasonal labourers. And isn't Bingham an occupational traveller. I mean, *what else*? Unlike the Romani who, whatever the varying national statuses of their claims, have undeniable ethnic identities, occupational travellers may not be marked this way, or may be comprised of mixes of many. Migrant farm labourers in the United States may be Mexican, but they may also be Jamaican, Guatemalan, Salvadorian, U.S. citizens of Hispanic origin, or U.S. citizens of other origins. They may have homes they intend to return to, homes they don't intend to return to, or none at all. The lack of even the doubtful visibility granted by ethnicity makes these travellers all the harder to map.

Finally there are the growing numbers of New or New Age Travellers who, originating in the 'crusty' hippy-cultures of the 1970s, lack even a 'decent' pedigree and merge, via free-party sound-systems like DiY and Spiral Tribe, into the world-wide free-tekno and rave circuits. How many New Travellers are there? No one has a clue but wandering around the UK alone – some in their third or fourth generation – may be as many as 40,000. We *do* know in 1992 that, 40,000 New Age Travellers, Gypsies, and ravers descended on the Castlemorton Commons for the week-long party that inspired passage of the Criminal Justice and Public Outdoor Order Act of 1994, an act that not only outlawed raves but vacated the requirement that local authorities had to provide campsites for Travellers of any stripe, New, Occupational, or Romani (Reynolds 1998).

Okay, these *are* embattled populations, but … *reclaim territory?*

They never had any territory to begin with which, I insist, doesn't mean they don't have a place. Furthermore, it's plain that very many of them *don't want* any territory. Like Ryan's, the place they occupy is anything but the sum of its habitats. Movement is what best distinguishes their place, that movement so anathemic to states with their fetishes for location, location, location.

If only these people would settle down!

Settlement has been the strategy adopted by nearly every state confronted with mobile populations like these: sedentarization together with what the Spaniards subduing the New World called *agrupación*, that is, not sedentarization alone, but sedentarization into aggregations large enough to efficiently control, tax, catechize, indenture, conscript, educate … It's what Israel's doing with the Bedouin right now. I suppose it goes without saying that they're severing the Bedouin from their flocks, evicting them from their tents (the Israelis bulldoze these), but they're also *aggregating* them to facilitate their control and exploitation as … *labour* (Meir 1997, Shoshan 2010: 380). The Jahalin Bedouin, for example, have not only been forcibly removed from the land they previously occupied where the Jewish settlement of Ma'ale Adumim now stands, but have been recruited to work as domestics *in* Ma'ale Adumim, and as labourers in its unceasing expansion.[8] No longer, if Israel has its way, shall cares, 'fold their tents, like the Arabs,/And as silently steal away' (Longfellow 1846: 80), but rather like the Jahalin, remain in their shipping containers next to the Jerusalem city dump which is where the State of Israel has settled them.

In this case, there *may be* territory to reclaim, though it remains unclear exactly what this might be or how to map it. Thus a Jahalin Bedouin, Abu Dahook, explained that when he and his family lived on *private property* in Anata and Abu Dis, 'No one bothered us. The people in our towns and villages are honorable

8 It's maybe beside the point, but the conclusion in 'Arab Jahalin: from the Nakba to the Wall' (Hunaiti 2008) that the Israeli actions amount to ethnic cleansing is, I conclude after four field visits, incontestable.

people. If the land is empty, what difference does it make to the owner whether we live on it? On the contrary, when we live on the property, we protect it. That's how people looked at it. None of them ever asked the Bedouin to leave the lands they lived on' (Audeh 2008).[9] So … *what territory are we supposed to map?* It's not all that different from trying to figure what territory to grant New Age Travellers in, say, England: *what? include the entire country?*

There are conflicting problems here. The first is the one I've just alluded to: New Age Travellers, Gypsies, Bedouin, they all live on the margins, in the interstices, and they have a propensity for fading into the background and disappearing. Pinning them to a map can only mean posting their potential ranges which, as I pointed out about Bingham, means the *entire* geographical area containing suitable habitats. Given that range maps post neither locations nor numbers, what exactly is the point? Besides that they're just plain redundant. Again, on the map include what? *The whole West Bank? The entire continent of Europe?*

But the other problem is worse, for in our world where *the mapped alone is assumed to exist* – especially when it comes to occupation and/or legitimation – exclusion from the map amounts to, or foretells, some sort of disappearance. Some find this advantageous – not everyone is eager to exist in the eyes of the state – but if you want to exist, as the Bedouin emphatically do, then being excluded from maps is a liability, one that in the Bedouin case dates to the Mandatory British resurvey of Palestine (Gavish 2005). If the titles established for the *fellahin* at that time prove less secure than the *fellahin* might have assumed – and we'll ignore the loss of their common *musha'* lands – no titles of any kind were established for the Bedouin since the British, with their sedentary prejudices, surveyed only … the *settled* parts of their Mandate. The resulting absence, exclusion, oversight, is what fuels so much Indigenous counter-mapping today: in a world where maps matter, better to be on it than not (Wood 2010: 129–42)!

The problem with this is obvious. It's the same one that confronts animals – bears, elephants – when they stray across a park's boundaries; that faces kids playing in the street who are supposed to be – damnit! – in the playground (Wood 1977); that bedevils anyone, actually, who wanders off the reservation. Once you've been pinned to the map, that's where you belong, and let's have none of this nonsense about 'Bedouin moving around wherever there is food and water for the herds' (as Abu Dahook put it). *No! No! No!* Being on the map – having a location – means being settled. Being settled and having an address are just different ways of saying the same thing. Both are but different faces of a single project of the state. Ultimately this is the problem with the map for Indigenous, especially for nomadic populations (including New Age Travellers and students and Gypsies and young unemployed university graduates[10]): the very best the map can do is to transform

9 Indeed ownership *per se* rarely seems to be an issue.

10 Here I'm especially thinking of China's 'ant tribe', so called because there are so many of them: young, recent graduates of universities unable to find work in the large cities

their places into *its locations*. When places can't be crammed into the tidy boxes of the state, maps are helpless.

Toward the Right to a Place in the World

There is a dawning recognition of this reality. Writing in 1999 as president of Australia's National Native Title Tribunal, Graeme Neate pointed out that surveyors need to understand that:

> The rights and interests of indigenous people in their traditional country will not necessarily accord with conventional legal notions of property;
>
> In some areas two or more groups of people may have mutually recognized traditional rights and interests;
>
> In some areas the boundaries of traditional estates may be clearly defined by reference to natural features, but elsewhere the boundaries are imprecise, permeable, and periodically negotiable.
>
> It may not be possible to plot traditional estates or significant sites by conventional cartographic means, or record them cadastrally. Rather than attempt to record such estates and sites by using cadastral boundaries, it may be better to note, by references to areas mapped for other purposes, which group has (either alone or with others) which traditional rights and interests. (Neate 1999)

Drawing on thinking like this Justice Robert French of the Federal Court of Australia, himself a former president of the National Native Title Tribunal, made a determination in a 2002 land-rights case that almost came to an appreciation of a people's *place* rather than its location. Accepting a *dish of sand* from the Martu Aboriginal people, French (2002) acknowledged that the 'symbolic gesture was a demonstration of the claimant's strongly-held belief in their ownership of their traditional territories'; and in his finding French came close to expressing regret about the necessity of reducing the claim to a map: 'Although the Court has to set boundaries in order to define the area of a native title determination, it is a fact that in the extremely arid regions of the Western desert, boundaries between Aboriginal groups are rarely clear cut. They are very open to human movement across them. Desert people define their connection to the land much more in terms of groups of sites, thinking of them as points in space not as areas with borders' (French 2002). French approvingly quoted from an anthropologist's conclusion that if the inhabitants recognized any sort of territorial boundaries then, due to the patchiness

---

where they're landing. We're talking hundreds of thousands of unemployed, young adults. Most of these seem to be aiming for stability but …

and unreliability of the rainfall, that these boundaries had to allow people to cross them freely; and French went on to determine that indeed there existed *concurrent native title rights* of both the Martu and Ngurrara in the desert in question. 'It is particularly encouraging', he concluded, 'that each of these groups, consistently with their traditional law and custom, is able to recognize the interests of the other in a common area of land' (French 2002).

Despite the decision's inclusion of an endless list of the lat/long coordinates fixing the area of the determination, French's recognition of concurrent title rights, his calling into question 'clear cut boundaries', phrases like 'thinking of them as points in space not as areas with borders', and the grudging quality of 'although the Court has to set boundaries', suggest a focus on place at the expense of location. As does, in a wholly different context involving Occupational Travellers, the decision of the Brisbane School of Distance Education to have teachers travel *with* fairground, circus and carnival workers rather than force their children to attend school in a fixed location. Instead of disrupting the students' families, the school itself became mobile; that is, rather than dissolving the children's place, the school conformed to it (Danaher, Danaher, Moriarty 2003: 166–7). An approach in the United Kingdom permits dual registration for Traveller students, holding open an absent student's place at one or another school while recording absences as excused.[11] Here it's less an individual school conforming to a Traveller's place than the system as a whole. Helping kids stay in any school was the UK's 1968 Caravan Sites Act which, by requiring local authorities to establish caravan sites 'for the use of gipsies and other persons of nomadic habit', saw the creation of 400 new caravan sites across the country (Caravan Sites Act 1968).

Of course this was undone by the 1994 Criminal Justice and Public Outdoor Order Act, but what I'm trolling for are hints, suggestions, models of things that have been or could be done to guarantee a person's *place in the world* against a minimal right to a *location on a map*. Of course the various rights enshrined in the United Nation's Universal Declaration of Human Rights and its Declaration of the Rights of Indigenous Peoples are fundamental, but without what Robbie McVeigh (1997: 7–25) has called an 'unambiguous acceptance of the right to travel', they come to little for people on the move, especially for people who, like Occupational and New Travellers, may fall into the category of what the UN act refers to as 'all other peoples' (United Nations 2007).[12]

11 See, for example, the London Borough of Havering's Traveller Education Support Service (TESS), though all Local Education Authorities have something along these lines. While there is some visiting of students and parents on privately owned sites, fairgrounds, and circuses, as well in private and Council accommodations, a great deal of focus is put on materials that can move with students rather than on movement of the teachers.

12 Words including 'nomad', 'nomadic', 'travel', 'traveller', 'mobile', and so on do not appear in the UN's native rights declaration. Of course, phrases like 'distinctive spiritual and material relationship with the lands they have traditionally owned or otherwise occupied or used' could be interpreted as including nomadism.

The right to travel – wow! hard to imagine in a world of borders – but it happens to imply … the right to park. I mean, *here's* a right we almost take for granted, all us people on the move who not only assume we'll be able to find a place to park, but that we have a right to it! We don't, of course, no matter the appearance; and on another scale it was precisely this lack of a right to park that the Caravan Sites Act addressed. A right to travel might also imply that government services are services for citizens, not just for sedentary citizens; and that this might mean *all* services that governments provide, including education and medical care, and here of course the Brisbane, UK, and other schooling initiatives are illustrative. Stir in the sorts of things Justice French was getting at in his decision concurrent title rights, vague boundaries, spaces conceived as points in space rather than areas with borders – and I think we may begin to have some of the dimensions of a place in the world.

Could we think about these the way we think about niches? As dimensions of a multidimensional space to which we could … *grant title?* Why title? Because title-holders seem privileged in ways the possessors of mere rights don't (*vide* the Bedouin); and because I'm not thinking about grazing *rights* or *rights* of access or passage, but about complicated *bundles* of such rights (including the right to travel) combined with accesses to services and maybe even chunks of land (the maps for these attached as codicils), and all these bundled up together, the way banks bundle mortgages up into securities which people then buy and sell. I'm not thinking about buying and selling place titles – though why not? – but about governments granting them the way they used to issue land grants, granting them to people who have *places* but don't particularly need *territory*, or need territory but only now and then, in some sort of periodic or rhythmic way. And, okay, the grant would come with a 'place title number' attached to it, a kind of licence – we do live in a system of states after all – but this licence would guarantee its holder the right to have school absences excused, or to park, or to cross a border, or to graze animals.

I don't know. It's just a thought.

Acknowledgements

This chapter was delivered as a keynote at the *Nomadic and Indigenous Spaces: Productions and Cognitions* conference, University of Leipzig, Leipzig, Germany, 3–4 February 2011. It was accompanied by 95 images which are available as a PowerPoint presentation from the author at his website: www.deniswood.net. The text itself was accompanied by copious interesting endnotes, jettisoned to conform with the publisher's house style. These are also available at the author's website. The brilliant editing of the chapter was carried out by Peter Koch and Angelika Brünecke, to whom all thanks are due!

References

Audeh, I. 2008. *A Constant Nakba for Palestine's Bedouin (Part 1). Electronic Intifada* [Online 7 July 2008]. Available at: http://electronicintifada.net/v2/article9665.shtml [accessed: 14 December 2010].

Bemelmans, L. 1934. *Hansi*. New York: Viking.

Bonner, J.T. 1969. *The Scales of Nature*. New York: Harper & Row.

Caravan Sites Act. 1968. *Caravan Sites Act* [Online]. Available at: http//:www.legislation.gov.uk/ukpga/1968/52 [accessed 4 January 2012].

Csete, A. 2006. Ethnicity, conflict, and the state in the early to mid-Qing: The Hainan highlands, 1644–1800, in *Empire at the Margins: Culture, Ethnicity, and Frontier in Early Modern China*, edited by P. Kyle Crossley, H.F. Siu and D.S. Sutton. Berkeley, CA: University of California Press, 229–52.

Danaher, P., Danaher, G. and Moriarty, B. 2003. Space invaders and pedagogical innovators: Regional educational understandings from Australian occupational travelers. *Journal of Research in Rural Education*, 18(3), 164–9.

French, R.J. 2002. *James on Behalf of the Martu People v Western Australia. FCA 1208* [Online, 27 September 2002, updated 6 July 2004]. Available at: http://www.austlii.eduau/cases/eth/federal_ct/2002/1208.html [accessed: 11 November 2011].

Gavish, D. 2005. *A Survey of Palestine Under the British Mandate, 1920–1948*. London: RoutledgeCurzon.

Hughes, S. 1979. *Surveyors and Statesmen: Land Measuring in Colonial Virginia*. Richmond: Virginia Surveyors' Foundation and the Virginia Association of Surveyors.

Hunaiti, H. 2008. *Arab Jahalin: From the Nakba to the Wall. Palestinian Grassroots Anti-Apartheid Campaign. Ramallah* [Online]. Available at: http://www.stopthewall.org/downloads/pdf/Jahalin-EN1.pdf [accessed: 14 December 2010].

Hutchinson, E. 1978. *An Introduction to Population Ecology*. New Haven, CT: Yale University Press.

Kain, R. and Baigent, E. 1992. *The Cadastral Map in the Service of the State: A History of Property Mapping*. Chicago, IL: University of Chicago Press.

Longfellow, H.W. 1846. *The Belfry of Bruges and Other Poems*. Cambridge: John Owen.

McVeigh, R. 1997. Theorising sedentarism: The roots of anti-nomadism, in *Gypsy Politics and Traveller Identity*, edited by T. Acton. Hatfield: University of Hertfordshire Press, 7–25.

Meir, A. 1997. *Nomadism Ends: The Israeli Bedouin of the Negev*. Boulder, CO: Westview Press.

Neate, G. 1999. *Mapping Landscapes of the Mind: A Cadastral Conundrum in the Native Title Era*. Paper to the UN-FIG Conference on Land Tenure and Cadastral Infrastructures for Sustainable Development, Melbourne, Australia,

25–27 October 1999 [Online]. Available at: http://www.fig.net/figun/sessions/session5/neate.pdf [accessed: 3 January 2012].

Nietschmann, B. 1995. Defending the Miskito Reefs with maps and GPS: Mapping with sail, scuba, and satellite. *Cultural Survival Quarterly*, 18(4), 34–7.

Reynolds, S. 1998. *Generation Ecstasy: Into the World of Techno and Rave Culture*. Boston, MA: Little, Brown and Company.

Ricklefs, R.E. 1990. *Ecology*. 3rd Edition. New York: W.H. Freeman.

Shapin, S. and Schaffer, S. 1985. *Leviathan and the Air-Pump*. Princeton, NJ: Princeton University Press.

Shoshan, M. 2010. *Atlas of the Conflict*. Rotterdam: 010 Publishers.

The London Borough of Havering. *Havering Traveller Education Support System* [Online]. Available at: http://www.havering.gov.uk/Lists/Related%20Media/Attachments/141/Havering_Traveller_education_service.pdf [accessed: 4 January 2012].

United Nations 2007. *United Nations Declaration on the Rights of Indigenous Peoples*. [Online]. Available at: http://www.un.org/esa/socdev/unpfii/en/declaration.html [accessed: 3 January 2012].

Wilford, J. 2000. *The Mapmakers: The Story of the Great Pioneers in Cartography from Antiquity to the Space Age*. Revised Edition. New York: Alfred A. Knopf.

Wood, D. 1977. Free the children! Down with playgrounds! *McGill Journal of Education*, 7 (2), 227–42.

Wood, D. 1982. Review of the book *The Human Experience of Space and Place* by A. Buttimer and D. Seamon. *Environment and Behavior*, 14(4), 503–6.

Wood, D. 2010. *Rethinking the Power of Maps*. New York: Guilford.

Wood, D. and Fels, J. 2008. *The Nature of Maps: Cartographic Constructions of the Natural World*. Chicago, IL: University of Chicago Press.

Chapter 3

From Nomadic to Mobile Space: A Theoretical Experiment (1976–2012)

Denis Retaillé

My empirical work about the encounter of nomadic and peasant people in eastern Niger during the last three decades of the twentieth century has led me to state and scrutinize categories and concepts used to describe a *crisis*. In the 1970s, there was widespread discussion about the end of nomadism in the Sahel and the critical situation of the region. Where was this *crisis* located? Perhaps in the categories and concepts themselves. This paper tries to show how they can be comprehended and why. Globalization has also been presented as a *crisis* due to the mobility and uncertainty it creates. Therefore the Sahelian example of mobile space could be a way to understand and illustrate what happens to the whole world when uncertainty becomes the rule. The mobile space theory I am going to present here must be understood in constant progress, based on a new geographical axiom. It begins like this: mobility comes first, before anchoring. This idea is in contradiction with a long geographical history based on sedentary power.

This narrative, a journey from experience to experiment, might be better understood orally, for orality is a more appropriate and useful knowledge pattern in the description of *mobile space*, fluid as words and not fixed on maps. This is precisely what I learned when doing fieldwork in nomadic spaces of the southern Sahara and Sahel. The experiment is built on the nomadic spatial experience; this is the focus of the article. More exactly, it is about the meaning of place: how does space *take place*? In nomadic space, each place is not only a stop but also a crisscrossing, a junction because movement comes first and represents the only possibility to meet someone or something, which is itself in motion. Thus, this meeting is like a narrative: the narrative of an encounter.

The Sahelian Experience[1]

My account starts in the 1970s, when I found myself in the Sahel and the southern Sahara during the most severe drought of the twentieth century. It was a time of

1 From 1976 to 1998, I published several papers, each corresponding to one of the many stages of this experience. The main ideas were summarized in 1989 (Retaillé 1989) and several figures in this chapter were first published in that text. I dedicated a chapter

death, but the narratives of ancestors and of their places were nonetheless all about life, as anywhere in Africa and worldwide. I learned it from nomadic people: their space firstly exists in memory and it can be expressed through words. However it never depends only on memory because space is also imagined, projected or seen as planned spaces, such as reserves, escape or survival spaces, or as a metaphor of solidarity to be created in times of hardship. Hence the narratives of places are superior to the map.

I was a geographer and I asked myself: could we draw a map only with words or, at least, with the forefinger pointing at the stages of a road? But, what kind of stages? Stages of motion. So I had to learn a new geography. My masters were nomadic people and, sometimes, sedentary ones, but the latter too were *mobile*. Orality is called upon to give a meaning that does not exist in material sites: sites have no memory, only people do. I will try to show why, introducing the concept of mobile space and *spontaneous geography*. It began like this (and back then I had to use a map).

The Drought and the Moving Frontline between Peasants and Herders

During the 1970s' drought in Africa, geographers and other scientists thought they were observing the end of nomadism, with the death of livestock and nomadic people sedentarizing (Monod 1975, Gallais 1976). Why? It was not the first drought, after all. Was it more severe, then? Not really, as far as we could assess. So, what was happening? Why were the previous strategies of escape no longer being used? The answer to these questions was in the maps, the maps drawn by sedentary people (geographers are sedentary in mind although they travel), the maps where certain lifestyles appear to coincide with the natural environment. Here: pastoralism, livestock and nomadism; over there: peasants and agriculture. The limit? An isohyet.[2] But, what kind of isohyet? The average one? Yes, of course. Yet the mere idea of average is nonsense here because from one year to the next, rainfalls are very irregular. The map is a document with fixed categories, among them the zones of production and the categories of sedentary and nomadic *vocations* (see below, Figure 3.4). In Niger, legal documents passed in 1961 draw this limit between agriculture and pastoralism.[3] It is total madness.

---

of a collective book in 1998 to the transfer of nomadic space concepts into other contexts (Retaillé 1998a); more recently I wrote two papers on this topic with Olivier Walther (Retaillé and Walther 2011, 2012). In addition, three papers were published about the 'African lessons' (Retaillé 1993, 1998b, 2004).

2 Isohyet: a line that connects points with the same amount of (average) precipitation (editors' note).

3 Laws 61-05 and 61-06 and decrees 61-150 and 61-254 of the Republic of Niger define the limit of agriculture and the zones of pastoral modernization. These legal documents can be found on http://www.case.ibimet.cnr.it/den/Documents/code_rural/cdrom/doc pdf/ (accessed: 1 July 2012).

So, during the 1970s and early 1980s geographers observed this famous limit (the isohyet) moving southward, although it was not really the case. Moreover, the limit could not possibly be a front line. From the point of view of a zonal and biogeographical definition, yes, the Sahel was moving indeed. The actual front line between nomadic and sedentary people did not correspond with the legal one. In fact, this limit was not a fixed border but rather a shifting zone.

As a first step, I modified the classical zonal model and introduced other features (see Figure 3.1). I did realize that during rainy years peasants pushed northward, trying to gain more agricultural land and that during dry years herders pushed southward, seeking pastures. If the Sahelian zone was to be defined by the encounter of farmers and herders, the zone had to be mobile. This insight was the first warning that something is fundamentally wrong with the axioms that we had been using.

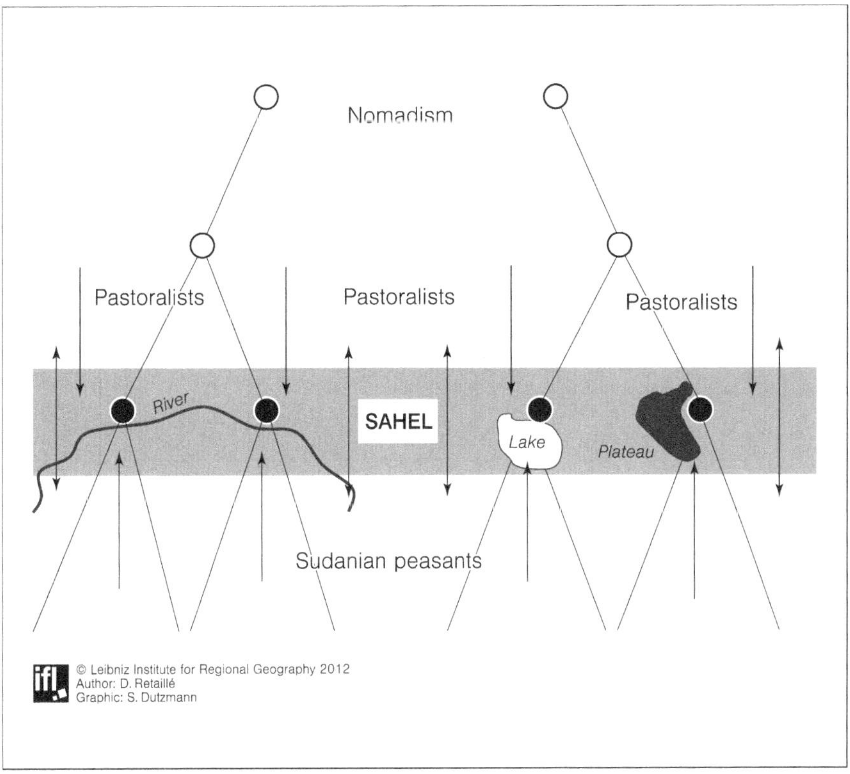

Figure 3.1 Encounters of nomads and peasants in the Sahel (a zonal model with a mobile front)

Source: Author, first published in 1989

Observing these movements more closely, I noticed that when farmers pushed northward, they were facing uncertainty. It was not possible to move forward and backward every year. Movements northward seemed to follow river valleys, lakes, foothills and cities, leaving large spaces between these sedentary areas to herders, spaces that were very useful during dry seasons or as zones of refuge during long periods of drought. More precisely, cities and areas of high density develop where roads meet valleys, lakes, mountains or plateaus. So, the Sahelian zone could be better understood as a line of – very old – cities rather than a bio-geographical domain in relation with a specific spatial organization. The next map expresses that concept (Map 3.1).

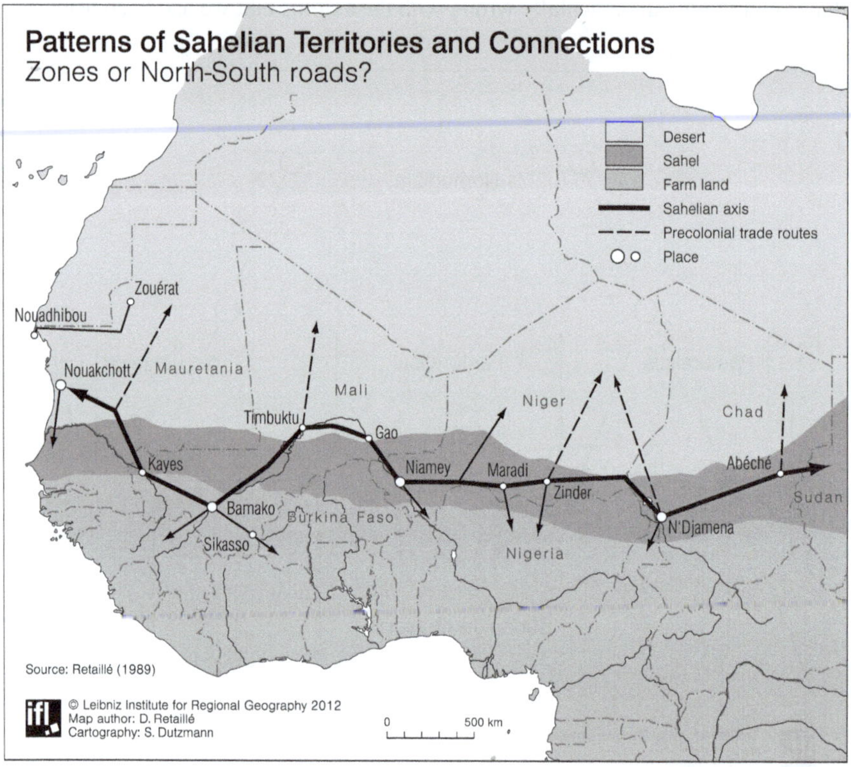

Map 3.1 Zones or North–South roads? Patterns of Sahelian territories and connections

From Roads to Areas: The Territories of States

One obvious thing here is how state territories have been delineated according to the same pattern. They are *based* in the Sudanian belt, which is a sedentary one, and the cities are linked by the West–East *colonial road* – the Sahelian axis – whereas

North–South roads leave the desert. Towards the north, the state territories fan out into the desert. However, it was more interesting to see how the multiple strings and bundles of the main nomadic routes actually connected zones of high density of sedentary people (Map 3.2).

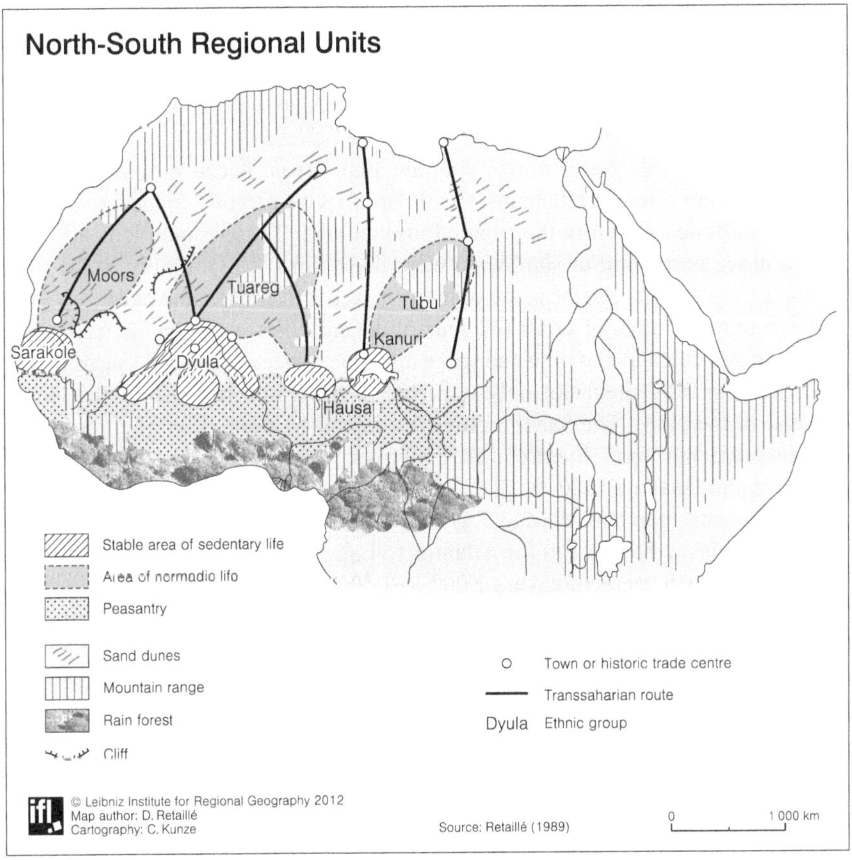

Map 3.2 North–South regional units

I found the metaphor of connection to be the best tool to reflect on this. Taking the possible movements of Sahelian people northward and southward into account, both nomadic and sedentary, I changed my scale of analysis and began to conduct very local fieldwork. The first attempt of mapping (Map 3.1 and 3.2) only partially showed the Sahelian spatial organization, as it concentrated on the regional scale (and not the infra-regional and local ones). In addition, it was still based on a geography of production zones, lifestyles, and vocation in accordance with a supposedly *average environment*. But fieldwork and life rhythms, rather than average numbers, maps and categories, draw another geography where city and market lines also moved.

Depending on climatic, economic or political circumstances, the main locations where nomadic and sedentary people could meet were the northern or southern ones. When colonial boundaries were drawn, for example, and, even more so, when national borders were recognized by the new African Union Organization in 1964, these main places moved and found themselves close to the new borders, along the new West–East national highways. What used to be the trail for colonization had now turned into the new states' *Unity* roads (Map 3.1). The desert became a marginal zone and a dead border itself; or so we believed.

Roads and trails were more important than areas, because people localize themselves along lines and chains of solidarity (with parents, friends, managers, patrons and clients) before they spread into an area. The preconditions allowing one to move northward or southward were these links. The mobility of the front line was not a frontier but the result of many mixed movements that occur regularly and not only in times of crisis (of course, during crisis, movements were more intensive and extensive in time and space and thus easier to observe). Surprisingly, extensions in movement not only occurred to the South but also to the North, because of the cities. The Sahel region was organized along lines of cities, which however did not form hierarchized networks; rather, they formed 'rhizomes' (Deleuze and Guattari 1980: 9–37).

The general map above (Map 3.1) shows the spatial arrangement from Dakar or Saint Louis on the shore of the Atlantic, to Lake Chad and the erstwhile Empire of Ouaddaï. Each segment covers 1,000 to 1,500 km, but the crossroads with the old north–south trails structure space more reliably. It was there that the informal market places appeared and then, they could move: to the North or to the South; to the West or to the East. Furthermore, we should remember how the main Trans-Saharan road moved from West to East during the millennium preceding colonization.

Changing Scale: A Village at the Edge of the Desert

These north-south trails are to be observed very attentively. This is what I did in East Niger along a road that was a bundle of trails linking Nigeria to Libya through Koutous (in the northern part of Gouré) and Kawar (Bilma). I stayed in Boultoum (1981–84), the last *sedentary* village and market before the great crossing of the desert. All kinds of people were here: nomadic people such as Tuareg, Tubbu, Fulani and peasants such as Kanuri and Hausa. The Tuareg, Tubbu and Fulani were actually peasants, whereas the last caravan I saw in 1984 was conducted by Kanuri, and the Hausa controlled the market and exchanges. Sometimes, Arabic people appeared from Chad and Libya. It was the time of civil war, and many weapons and orders passed through this spot. Boultoum was also a stock exchange, and although nobody (or only very few people) seemed to live there, it was a central place. How could this be?

I thought the answer could be found in the very idea of movement and I tried to catch this movement leaving behind the too simplistic concept of zonal *shifting*. I learned from experience that the same place could *take place* in any site elsewhere, be it to the North or to the South. I also learned that the trail could be far to the

Figure 3.2 The market place of Boultoum

Source: Photograph taken by Denis Retaillé in 1981

West or far to the East. I was not yet ready to understand that space itself was mobile, and yet it is exactly this insight that later occurred to me. Uncertainty was not only a matter of climate and rainfall. I learned that everything is uncertain, and first and foremost, it is uncertain whom one will meet. This is why mobility is a survival condition. Therefore everyone gathered here at the same site to meet everyone else. Here, one could always find at least one member of one's own tribe to feel like being at home. This is how home is everywhere, through lineage, tribe, ethnic identity and so on. I believe that this is the actual basis of movement.

Locally, another movement could be observed. During quiet periods, human settlements tended to spread while in times of crisis they tended to gather along the main road. Thus, the main road would look different each time.

It is possible to transport this sort of *space and place* where everybody can move for any reason: drought, war or economic opportunity. Along the north–south line, parents, friends, patrons or clients will be found. Along the line, one is at home. Home is mobile according to circumstances: this is the basis of spatial figuration.

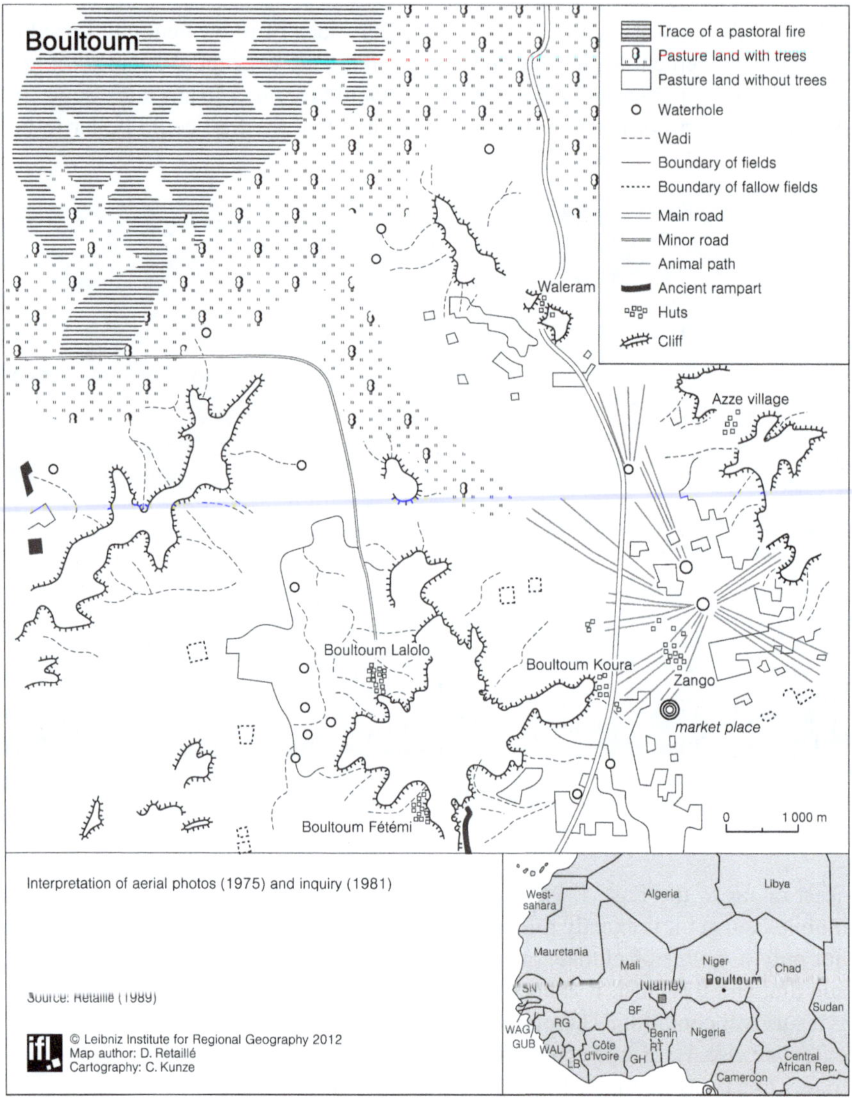

Map 3.3 Map of Boultoum (based on an aerial picture)

Early Lessons About the Idea of Place

Even though this model expresses the idea of Sahelian mobility to some extent, I no longer find it entirely fitting. When the sketch was published in 1989, I was still trying to preserve the zonal model and distinguish people using a lifestyle typology. Yet, not only space is mobile: so are social and even ethnic conditions

(Urry 2000). The social and spatial patterns need to be joined, but most of the geographers have not yet admitted the mobility of space itself.

Last but not least, another major difficulty lay in the definition and demarcation of ethnic groups. Nomadic and sedentary peoples were radically distinct in terms of ethnic identities: yet, similar to the geographical limits drawn on maps, the validity of these cultural and ethnic boundaries was overestimated. I had to admit that a majority of ethnically nomadic people were sedentary, and that among the ethnically sedentary groups many individuals were mobile. Furthermore, mobility appeared to be the social sign of a relatively high status and well-being, both for nomadic and sedentary people. Power lay in the control of movement (Deleuze and Guattari 1980). *Nomadic* and *sedentary* were no longer the relevant criteria. This was the second warning, the second hint about the existence of mobile space, which could be described by the dynamics of competence and power for both sedentary and nomadic people. This space of capacity fluctuates and its points of references move.

The main lesson emerging from this insight was about place. What is a place in a nomadic (or now a mobile) space? What is a place, if places can move? Boultoum, this ghost village, was indeed a central place. I then understood that it was necessary to distinguish very clearly between place and site, place and locality. For me, place could no longer be defined as a localized point with specific and definitive characteristics, nor by a singular name: one place can have several names (Boultoum, Boultoum Koura, Waleram, Zango, Fetemi), depending on different references. Only one property could define it: a place is where the distance between people is zero in terms of space (copresence), and a place can be moved to another location, if necessary, to preserve the condition of absence of distance.

Transferring the Idea of Mobile Space to the Globalized World: An Experiment[4]

Taking this definition of place as a starting point, it seemed that a good way to translate the nomadic space concept into our contemporary *modern* or *hypermodern* condition could be to look more closely at globalization. I am of course well aware of the dangers of metaphors when comparing the hyper-mobility of the 'rich and wealthy' with the know-how and strategies of nomads who have to try hard to make ends meet (Amselle 2010, 2011). I only attempt to adapt the hypothesis of place in a nomadic space to the idea of place in a globalized world as movement comes first in both cases (by hypothesis).

4 The opportunity to experiment occurred between 1990 and 1997, when I taught Geopolitics at the Paris Institute for Political Sciences (Sciences-Po).

The Sahel: Conventional vs Spontaneous Geographies

The concepts related to nomadic space cannot very easily be transferred to the globalized world (Retaillé 1998a). Neither the economic nor the technical basis refers to the same level and, perhaps more importantly, the territorial state is nowadays the dominant form of state in the world. It is important to recognize that in the Sahel the states that were created after the colonial period altered the regional spatial structure, trying to designate specialized agricultural production zones according to climatic zones. Regional planning, politics of development, technical assistance… all these agencies share the same model which understands the Sahel as a region comprising various zones, each one with a specific economic purpose or vocation. Only the main cities, usually one for each state, attract all the wealth; reversely, in case of disaster, only the states can bring remedy to the population. In the Sahel, however, they could not because of their weakness. Their inhabitants progressively lost their spatial know-how and their strategies of escapes, and therefore the last drought was a tragedy. Politics to increase production level and regional specialization were a risk factor. It was not a paradox, only a geographical mistake. Before proclaiming *crisis* and *fate*, we have to check our own models and concepts. That was the lesson I learnt from the Sahel drought experience. So when the world is presented as being in crisis because of globalization, my question is: do our explanatory tools fit?

In the Sahel *zone*, ten years after the great drought, the year 1984 was very dry. I observed how the population then rediscovered the north–south spatial solidarity and mobility. I later also observed how, from the early 1990s, religious associations and underground trade started following the same spatial organization. This is what I call *spontaneous geography*, in opposition to the academic one. And although geography proved to be *real* as recent events in the Sahel have shown,[5] academic geography has failed to explain it (compare Figure 3.3 with Figure 3.4, see also Walther and Retaillé 2010).

'Y': The Properties of Place

The next phase of my theoretical experiment was developed within a collaborative work about globalization, which resulted in one of the first books published in French dealing with this issue: *Le Monde, espaces et systèmes* (Durand, Lévy and Retaillé 1992). My nomadic place concept revealed itself to be very useful at the time. One has to remember a place is neither a site nor a localized point (a locality with a name) but it is where events take place, amounting to the actual content of sites and localities or revitalising them. Encounters contribute to this actuality by which a site or a locality is made a lively place. *Taking place* is a charming metaphor in French

5 Exemplary for the importance of this 'spontaneous' geography are the routes used by people from sub-Saharan Africa trying to migrate to the European Union and the network of connections within Al-Qaeda in the Islamic Maghreb (AQMI).

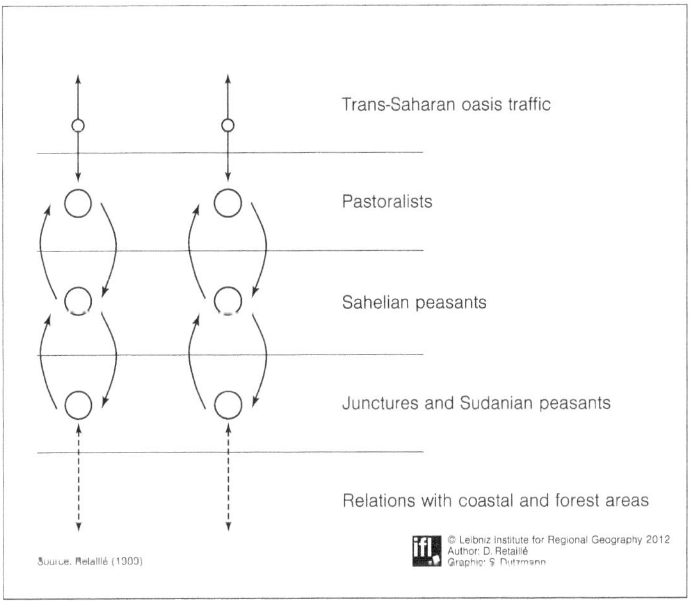

Figure 3.3 The Sahelian model of spatial structure on a large scale

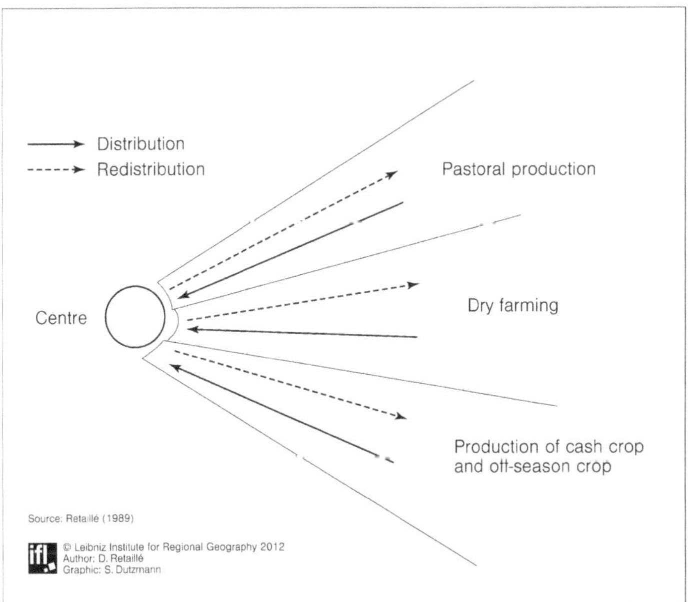

Figure 3.4 A development model of zones of production and vocations – which now appears to be a major conceptual mistake

as it translates as *faire lieu* and not *prendre lieu*. There is a one-letter word in French that expresses place: *y* (as in *il y a* or *il y est*) (Berque 2000). This is the property of the place: distance is zero, whatever type of measurement you use and *y* says that it is not exactly *here*, nor is it *there* because *here* and *there* point out a previous site or previous location (*ici, là*), whereas *y* is the abstraction of an encounter.

It means that if you can hear a voice in a room, then the room is a place. Referring to the academic institution, the building where we had our conference is a place. Through the different rooms and buildings, streets and stations we used together or alone during the three days of the conference, Leipzig became a place for us and for others we met, and so on. Place is not defined by any scale a priori; it all depends on connectivity. Germany is a place, Europe is a place and, I would add: the world itself is a place – the place of life, the place of humanity, the place of the global system I try to describe below. If different routes coming together are connected by some form of *salience*: then there is a place. Another salience creates another place – the same or not exactly the same. That is what we can understand by following nomadic people: the salience shapes the form of local association and the level of identity: ethnic, political, economic, *global*, or any other level of identity.

Four Levels of Salience

To keep things simple and to qualify any type of place in the world, I have defined four levels of salience depending on the kind of link that creates the style of unity and identity of the place. The four levels of salience should not be mistaken as a functionalist typification of place; rather, they describe the formation of a dynamic system.

- Cultural salience (in French: *accroche culturelle*), when distance is made non-existent by cultural mobilization as with iconic places for example;
- (Geo-)political salience (*accroche géopolitique*), when power gains strength by affirming its mobilizing force, as for example with places of strategic importance in a military sense;
- Economic salience (*accroche économique*), when social links are mainly based on calculating economic growth, as with industrial or financial districts;
- World salience (*mondialité*), when public attention works on the scale, as with world heritage sites, for example, or with global environmental change, or with global slogans promoting peace, human rights, sustainability, and so forth.

In each and every place in the world the four levels come together, each to a greater or lesser extent. So places have different profiles and these can change. Places *take place* around sites or localities through a mixture of habitualization, economic opportunity, and inherited symbolic force. Yet in the same site or locality, different types of places can take place with different profiles as the profiles themselves change in time. In Figure 3.5, profile #2 is that of a locality characterized by its economic salience. Profile #3 would apply to the same locality if cultural and world salience grow in importance, for example in the case of industrial cities converted into creative cities, with old industrial buildings and warehouses being used for new companies. And profile #1 would describe the same place after an economic crisis before it turns into world cultural heritage; that is, at a time when the city's history consists of the personal memories of the city's inhabitants but does not build up to a coherent, socially shared imaginary.

Moreover, the same profile can apply to a place in several sites or localities at the same time and for the same reason. This type of network, without hierarchy, is what Deleuze and Guattari (1980) named 'rhizome' in their famous book *Mille Plateaux*.

Mobile Places: The Beginning of a Theoretical Alternative

To see how the concept of mobile space relates to the definition of place in the globalized space, let us now summarize several points about and characteristics of nomadic (mobile) places. First, we can see the importance of the event (salience) to organize routes (be they material or not); second, one has the possibility to choose from several sites or localities to *take place*; third, of course, there is the possibility to have several places on the same site or locality. This is an important feature to understand the concept of globalized places. Moreover, all places are more or less globalized nowadays. This is the beginning of the theoretical alternative I later propose. This alternative draws on relational geography and it is the only one I know which adopts a radical renewal of the conceptual tools due to the influence of nomadic lessons. To make this point clear, I include two longer quotations from a recently published working paper (Retaillé and Walther 2012: 2–3):

> Since the pioneer work of Gottmann (1952), several generations of geographers have examined how the tension between mobility and place has been affected by changes in the international political economy and cultural politics. The intensification of global flows throughout the 1980s and 1990s contributed to a growing interest in the 'hostility' between territorial and relational modes of space organization and to a passionate debate within geography, fuelled notably by O'Brien's (1992) provocative theory on the end of geography, Ohmae's (1990) theses on a borderless world, and Castells's (1996a) seminal work on the network society.
>
> In recent years, debates on the relationship between territories and networks seemed to have abated: it is now acknowledged that places have not gradually disappeared and been replaced by a space of flows. On the contrary, it seems that networks – and the movement that goes with them – have emerged as the prime, but not exclusive, force of organization of geographical space. Most commentators agree that networks cannot be understood without reference to their territorial background. The main issue being not so much whether the

> space of flows will gradually supplant the space of places, but how these two types of space can coexist in a globalized world and be simultaneously analyzed (Castells 1999, Taylor 2007).
>
> ...
>
> [W]e argue that despite an overwhelming number of studies on flows and networks, most of contemporary geography remains strongly influenced by a 'sedentary' vision of space in which mobility is primarily seen as movement between fixed locations and territories as the key analytical units of geographic research. Following Amin (2002: 387), for whom "the language of spatial change remains that of assuming organization along scalar and territorial lines", we call for a revision of the way geographers identify their objects.

Places are mobile, short-lived, ephemeral and fugitive, they reappear according to circumstances, unchanged or altered. Movement is their first quality. Movement is required before any anchoring can happen, should it happen at all.

Movement First: The Need to Reverse the Geographical Axiom

Several complex questions, however, remained. Here, I shall only deal with one: what can be said about space? Once we accept the above definition of place that emerges from the nomadic experience, how can space be represented? Here I will not call upon the nomadic metaphor but rather discuss directly the conceptual tools and the model that allow for depicting the world in motion. Once again, this thought is close to Castells's work about spaces of flows as they have become the dominant form of space without the constraints of localities; also close to Taylor and his central flow theory (flows make places), reversing Christallers's Central Place Theory; and close to the relational turn in geography too (Allen 2003, Yeung 2005), which considers that power lies in the various relations between actors.

The four levels of salience I described earlier are analytic, not ontological. The age-old geographers' rule claimed: only one form for all phenomena, trying to fit the different instances into a *Procrustean bed.* So we are used to drawing our maps by choosing from two conventional frames of representation: the physical globe or the plane as an abstraction of our sphere. But, what about the spaces of humanity full of concerns, representations, choices and voices?

What about the Interrelation of Places and Space?

Following the definition of place I suggested before, and the four analytic levels of salience, I started to think that space is neither the sum nor a superposition of spaces but an integral line. I drew back to the mobile circumstances of places, integrating various levels of motives.

I speculate that the curves drawn in Figure 3.5 are integral lines, although I am not sure it is possible to determine the objective measures of each place because of their perpetual variations: it would be like stopping the movement, freezing it in time, as if in a snapshot. However, I am sure that we cannot simply put the different levels on top of each other and look at them as if they were a transparent map. That would be a sedentary and parochial vision. Place is not the mark of a tube from sky to earth or from earth to sky, as Aristotle (in his *Physics* IV, 4, 211) tried to convince us, 'the boundary of the containing body'. If movement comes first, we need a space of movement to describe it. This is the mobile space.

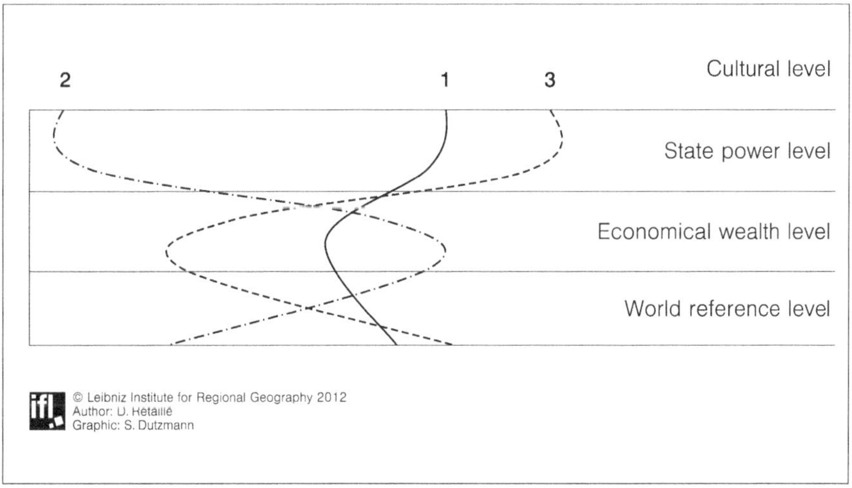

Figure 3.5 Profiles of salience that characterize a place

Source: Author, first published in 2012

Based on the new definition of place, we can create a space of representation (Lefebvre 1974) such as that in the model below (Figure 3.6). Integral space (the global one?) is the successive integration of each analytic level (cultural, geopolitical and economic salience), producing different types of space: the space of war (*espace de la guerre*) between cultural and state level; the space of order (*espace de l'ordre*) between states and international economic networks; the space of contract (*espace du contrat*) between the globalized world and world identity. And *vice versa*. Here, the ideal line of the world as a place would be the diagonal: identity would become less essential, the state would exist only as an instrument, economy would create welfare instead of profit, the world would become a project for all humanity.

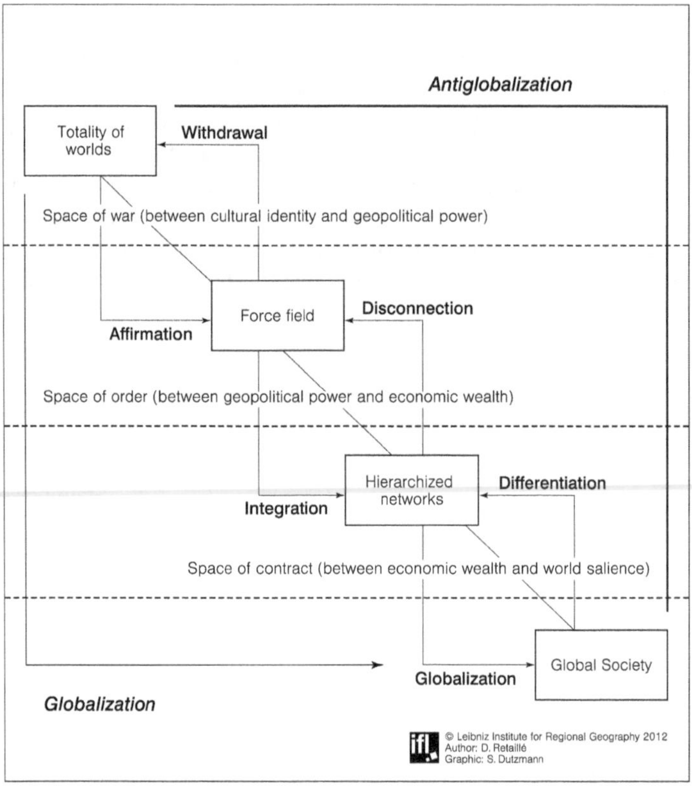

Figure 3.6 Mobile world space structure (a model, to be completed)

Source: First published in Durand, Lévy and Retaillé (1992), modified by the author

It is very difficult to formalize such a profile because of the numerous retroactions[6] in the process. The space of war not only represents the space of conflict space between cultural identities and the rule of the abstract state. The space of order is not only the space of well-known interests. The space of contract is not only the space of the globalized world. At the globalized scale, some activist movements call for a return to local identity. State financial support plays a role in both the space of order and the space of war, and so on. Many retroactions alter the situations. That is exactly why I believe nomadic space is a good starting point to try to understand the mobile space of the world.

These retroactions are only intuitive, but they all lead to the same conclusion: movement explains situations. We have to put movement first in our models, before defining situations and spaces. As movement appears to be a complex system of

6 Retroaction is used here and below as a shorthand for reciprocal action, or feedback process (editors' note).

retroactions, choices, and influences at the heart of the very definition of place, the nomadic example is again very useful.

Let us look at the system, which is very simplified here. Each link has to be identified and measured. How could this be done? As I worked out this model in 1990–92 (with my co-authors M-F. Durand and J. Lévy) I had to return to the nomadic hypothesis. If space is mobile, as I had learned in the company of nomadic people, it is quite impossible to use this model to conduct any kind of measurement. It could be used to recognize the various types of integral spaces, but these integral spaces did not give us any information about their (fixed) substances. In Figure 3.6, mobility dynamics are described on the left side: the creation of a *field of force*, the connection of economic networks organized into a hierarchy, and globalization. On the right side, retroactions are defined in sedentary terms. I argue that spatial analysis, just like classical geography, is unable to understand movement in a different way: movement in such geographies can only mean moving from one fixed place to another, and these places seem to depend on retroactions, not movement. In other words, the characteristics of places are defined first, whereas movement is considered an adaptation: the Central Place Theory is a perfect example of such reversed thought.

A Preliminary Conclusion: The Paradigm of Mobile Space

The dominant geographic paradigm and its axiom, since the 1960s, is what I call the *sedentary paradigm* (Figure 3.7). It can be synthesized as follows: features are anchored in the ground and maps structure them through only three types of elements: points, lines and areas. Three conditions explain both the differentiation and the organization of the world. Every scale is depicted alike. In fact, this is a structural scheme. It relies on three axiomatic precepts:

- Space is differentiated;
- In a place there is only one thing;
- Two places or two things are necessarily separated (there is a distance).

These three tenets come from *retroaction* as a response to movement.

The nomadic definition of place presented above requires an inversion. And this definition is the starting point of another axiom. If movement comes first, the features of any place are part of it, and not in pre-existent places. So I suggest another paradigm – a mobile space paradigm – one that is better adapted to a geography where space itself is mobile (Figure 3.7). And perhaps this paradigm was just hidden, invisible to the eye of the sedentary beholder. For how could a place take place without a movement first? Did it grow from the earth? Or did it fall from the sky? Even mythological foundations suggest some kind of movement.

Firstly, note how processes here lie in the movement itself, and not outside of it, as a legitimate cause and explanation. Secondly, according to the geographical

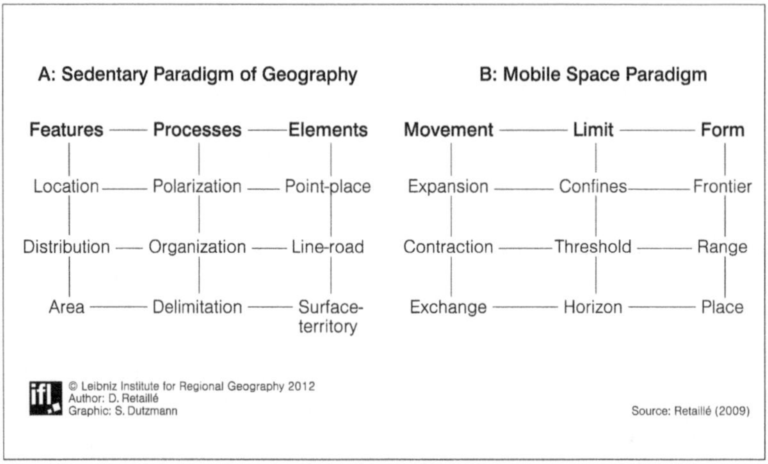

Figure 3.7 Space paradigms

Source: First published in French (Retaillé 2009), published in English (Retaillé and Walther 2011)

habit to cut the world into maps, this paradigm distinguishes three types of limits depending on the type of movement. And thirdly, geographical frames can be identified without using a map. Only words are needed.

This mobile-space paradigm is based on movement. It allows us to differentiate various levels and types of levels, especially in terms of unequal mobility.

- Scattering (divergence of flows, dispersion) can mean expansion or escape: it can be positive or negative;
- Gathering (convergence of flows, polarization) can mean meeting or concentration, struggle over resources, competition: it can be positive or negative;
- Exchange (crisscrossing of flows) can be equal or unequal: it can be positive or negative.

So now we can begin measuring: that is my main difficulty to translate this theoretical framework into empirical work.

This is still work in progress. I am convinced that the limits and their forms take centre stage. From the nomadic experience, I retain the last line of Figure 3.7. In a mobile space, places are created where any type of crisscrossing and exchange happens and it has no limit except one, the horizon (if the world were flat, it would have an outer boundary). There is no centre anymore, only centrality. And this centrality itself is mobile, creating multiple flows. I do not oppose a space of places with a space of flows as relational geographers do. Places come with flows: for example cities are flows in essence. The horizon is a limit without borders, neither to the inside nor to the outside. Once achieved, this mobile space is really the space of globalization. Its places are mobile and space is a new contractual space constantly being negotiated, no longer the space of *social contract* which was a philosophical and political fiction implicitly based on the assumption that space as territory is limited.

I also retain the first line of Figure 3.7. Unequal mobility puts some areas out of range. The limit is like edges, it may push or pull. Frontiers are mobile.

The middle line of Figure 3.7 – gathering – is more problematic for me, however, because it seems to be the description of fixed space precisely as stated by the sedentary paradigm. Does mobile space include fixed space? Which place is it in? Is fixed space the base of mobile space or rather a resistance to movements of globalization? If it is, this resistance shows a way. The control of movement originally serves as a base for power. The one who controls this space of mobility wins the land, wins the ability to exploit it in a sedentary pattern.

Postscript

This is the preliminary end of my search for a new theoretical approach to space and places. Having been asked about the fate of the people that initiated my search, I must admit that I do not know what Boultoum is like, today. Throughout that region, AQMI, some Tuareg people, the military junta of Mali and Pan-African forces, represented by CEDEAO,[7] are fighting for power using mobile space, on the one hand, and fixed space, on the other. Those who command the roads command the places without the necessity of covering the areas; such command places can move; in fact, they must move. Therefore it is so difficult to understand, deal with and solve this conflict. In early April 2012, part of the Tuaregs helped AQMI in their military campaign, showing them the roads and the manner to *take place* in the mobile space. Shortly after that, however, AQMI evicted these Tuaregs. While AQMI simply tried to gain control over roads and places, the Tuaregs wanted something different: they wanted to get hold of the territory, paradoxically following a sedentary approach to space. This leaves us with the question: Have they forgotten their own knowledge?
Bordeaux, April 2012. To be continued.

Acknowledgements

Thanks to Béatrice Collignon, to Joachim Otto Habeck and to Peter Koch who revised this chapter.

7 CEDEAO: Communauté économique des Etats de l'Afrique de l'Ouest (Economic Community of West African States, ECOWAS) (editors' note).

References

Allen, J. 2003. *Lost Geographies of Power*. London: Blackwell.
Amin, S. 2002. Spatialities of globalization. *Environment and Planning A*, 34(3), 385–99.
Amselle, J.-L. 2010. *Rétrovolutions: Essai sur les primitivismes contemporains*. Paris: Stock.
Amselle, J.-L. 2011. Méfions-nous de l'idéologie du nomadisme. *Le Monde* (24 June) [Online]. Available at: http://www.lemonde.fr/idees/article/2011/06/24/mefions-nous-de-l-ideologie-du-nomadisme_1540133_3232.html [accessed: 1 July 2012].
Berque, A. 2000. *Ecoumène: Introduction à l'étude des milieux humains*. Paris: Belin.
Castells, M. 1996. *The Information Age: Economy, Society and Culture, vol. 1: The Rise of the Network Society*. Oxford: Blackwell.
Castells, M. 1999. Grassrooting the space of flows. *Urban Geography*, 20(4), 294–302.
Deleuze, G. and Guattari, F. 1980. *Mille plateaux*. Paris: Editions de Minuit.
Durand, M.-F., Lévy, J. and Retaillé, D. 1992. *Le monde: Espaces et systèmes*. Paris: Presses de la Fondation Nationale des Sciences Politiques et Dalloz.
Gallais, J. (ed.) 1976. *Stratégies pastorales et agricoles des sahéliens durant la sécheresse 1969–1974*. Bordeaux: CNRS-CEGET.
Gottmann, J. 1952. *La politique des Etats et leur géographie*. Paris: Armand Colin.
Lefebvre, H. 1974. *La production de l'espace*. Paris: Anthropos.
Monod, T. (ed.) 1975. *Pastoralism in Tropical Africa*. London: Published for the International African Institute by Oxford University Press.
O'Brien, R. 1992. *Global Financial Integration: The End of Geography*. New York: Council on Foreign Relations Press.
Ohmae, K. 1990. *The Borderless World: Power and Strategies in the Interlinked Economy*. New York: Harper Business.
Retaillé, D. 1989. Comment lire le contact Sahara-Sahel, in *De l'Atlantique à l'Ennedi*, Abidjan: Centre Culturel Français, 19–33.
Retaillé, D. 1993. Afrique: Le besoin de parler autrement qu'en surface. *EspacesTemps*, 51–2, 52–63.
Retaillé, D. 1998a. Concepts du nomadisme et nomadisation des concepts, in *La planète nomade*, edited by R. Knafou. Paris: Belin, 37–58.
Retaillé, D. 1998b. Fantasmes et parcours africains. *L'Information Géographique*, 2, 51–65.
Retaillé, D. 2004. Afrique, oralité, géographie spontanée. *Bulletin de la Société de Géographie de Liège*, 44, 5–25.
Retaillé, D. 2009. Malaise dans la géographie: L'espace est mobile, in *Territoires, territorialités, territorialisation*, edited by M. Vanier. Rennes: Presses Universitaires de Rennes, 97–114.
Retaillé, D. 2012. *Les lieux de la mondialisation*. Paris: Le Cavalier Bleu.

Retaillé, D. and Walther, O. 2011. Spaces of uncertainty: A model of mobile space in the Sahel. *Singapore Journal of Tropical Geography*, 32(1), 85–101.

Retaillé, D. and Walther, O. 2012. *New Ways of Conceptualizing Space and Mobility: Lessons from the Sahel to the Globalized World*, Centre d'Etudes de Populations, de Pauvreté et de Politiques Socio-Economiques/International Networks for Studies in Technology, Environment, Alternatives, Development (CEPS/INSTEAD), Working Paper no. 2012-24 [Online]. Available at: http://www.ceps.lu/publi_viewer.cfm?tmp=1853 [accessed: 1 July 2012].

Taylor, P. J. 2007. Space and sustainability: An exploratory essay on the production of social spaces through city-work. *Geographical Journal*, 173(3), 197–206.

Urry, J. 2000. *Sociology beyond Societies: Mobilities for the Twenty-First Century*. London: Routlege.

Walther, O. and Retaillé, D. 2010. *Sahara or Sahel? The Fuzzy Geography of Terrorism in West Africa*, Centre d'Etudes de Populations, de Pauvreté et de Politiques Socio-Economiques/International Networks for Studies in Technology, Environment, Alternatives, Development (CEPS/INSTEAD), Working Paper no. 2010-35 [Online]. Available at: http://www.ceps.lu/publi_viewer.cfm?tmp=1577 [accessed: 1 July 2012].

Yeung, H.W. 2005. Rethinking relational economic geography. *Transactions of the Institute of British Geographers*, 30(1), 37–51.

Chapter 4

Where is Indigenous? Legal Productions of Indigenous Space in the Russian North

Gail Fondahl

Scholars have given attention to legal productions of *who is indigenous* – how the state[1] has produced and reproduced the category of indigeneity through legal enactments – including in the Russian North (for example, Donahoe et al. 2008, Slezkine 1994, Sokolovski 2003). Such productions of indigeneity usually have a strong spatial component, which law also serves to regulate – *where is indigenous.* Through the entangled legal constitution of *who is indigenous* and *where is indigenous* states have attempted to territorialize and re-territorialize[2] indigenous peoples within state bounds, simultaneously producing *indigenous spaces*, spaces that are not exclusively indigenous and spaces that exclude indigenous people. Law often also serves to legitimate such spaces, in the minds of at least a significant part of a state's citizenry.

Since the fall of the Soviet Union, the Russian Federation has pursued a re-territorialization project (sometimes glossed as 'land reform') on many fronts.[3] By state *re-territorialization* I refer to strategies used by the Russian Federation in the post-Soviet period to exercise power over its territory, through the partitioning of this territory and the controlled allocation of rights over these territorial units to groups and individuals, including in ways that are different from those pursued by the Soviet Union. A move from state socialism has required both changes in who has the authority to exercise power over territory (land and resources) in the Russian Federation, and the limits of that authority. Some lands and resources have been privatized, others reassigned from state organizations to private concerns, with varying degrees of rights to use, disposition and alienation. For instance, we see the lands of former state farms reallocated to peasant farming enterprises, agricultural production cooperatives, limited liability companies, joint

1 For brevity's sake I have anthropomorphized the *state* in this chapter. It is of course ultimately individuals who visualize spaces, write laws, interpret them, and use space. Greater attention needs to be paid to the role of various specific actors and actor groups (for example, lawyers) in the *state's* production of space (see Martin, Scherr and City 2010).

2 Territorialization differs from spatialization in that territoriality refers to the 'attempt to control people, phenomenon or relationships by delimiting and asserting control over space ... territorialization may entail the enforcement of spatialization' (Blomley 2010: 208).

3 Most land remains state property.

stock companies, etc. The re-territorialization project, part of a move toward a (quasi-)capitalist mode of production, has been expressed in, and facilitated by, the evolving legislation on property rights and access to resources.

This re-territorialization project includes and affects Russia's Northern indigenous population, in that the state has detailed new territorial visions and territorial units specifically for these peoples, including through legislation. This chapter addresses what kinds of *indigenous space* the Russian Federation is discursively producing via its federal legislation. It considers the rationales of justice and certainty as impulses guiding the evolving legal discourses of indigenous space. It attends only to the legal (legislative) production[4] of indigenous space at the federal level. Legal productions of indigenous space at sub-federal level have received modest attention, but deserve further inquiry, as does the production of indigenous space by means other than law.[5]

Producing Space by Law

> Legal decisions shape, demarcate and mould human geographies and social space. (Kedar 2006: 407)

I draw on Henri Lefebvre (1991), who asserts that space is not an *a priori* category. Rather, each society produces and continually reproduces its space, through political, economic, cultural and social practices and enactments (see also Blomley 2010). These practices may be discursive or material. Discursive practices include, for instance, investing meaning in a space by naming it, describing it. A space may be identified as a park or as an Indian reserve, or as Crown Land – in texts, in speech, in thought, in assumptions and narratives about land and property rights. Lefebvre refers to the 'representations of space' – 'conceptualized space, the space of scientists planners, urbanists, technocratic subdivides and social engineers' (1991: 38). We can refer to this as the *conceptualization* of space.[6] While Lefebvre does not explicitly include the drafters of law, these professionals can also be implicated as key players in statist conceptualizations and representations of space.

Material practices involve physical interactions with space, such as through embodied activities, maps, and infrastructure. The space of a park may be enacted materially through the construction of trails, the introduction of interpretive signage, the physical obstruction of motor vehicle access, and visits by hikers

4 The term *legal production* of indigenous space does not necessarily equate with the *lawful production* of such. Russian practices in this arena have often contravened international law.

5 For discussions of how some of the legal reforms have played out see, for instance, Donahoe (2006), Habeck (2005), Sirina (2005) and Stammler (2005).

6 Elsewhere others and I have used the term 'visualization' (Morris 1999, Morris and Fondahl 2002).

and picnickers. We can refer to this simply as *use*.There is also a middle ground between the discursive and material practices, where conceptualizations are translated into use. Lefebvre uses terms such as 'planning' and 'surveillance': I favour the term *administration* (Morris 1999, Morris and Fondahl 2002). Administration involves the organization, allocation and regulation of space. The space of a park is produced administratively through regulating what material practices can and cannot be carried out therein: for example, rules prohibiting motor traffic or allowing overnight camping.

States (state actors) produce and reproduce space through iterative conceptualizations, administration and uses. Law, as a narrative of state power, plays an especially important role in the state production of space. It naturalizes and legitimizes certain uses of space as 'appropriate'. It dictates access and exclusion. '[M]ost legal constructions are not ways of description but rather of prescription, of imagining *the possible, the probable and the desired*' (von Benda-Beckmann, von Benda-Beckmann and Griffiths 2009: 19, italics in original). The role of law in constituting space, as well as being constituted by it, has recently received increasing attention among geographers and anthropologists.[7]

Of course the customary law of indigenous peoples plays a similar role in producing space, in imagining and providing the 'rules of the game' for the desired relationships between individuals and groups in their territorializations. However, while deserving of much greater attention than they have received to date, indigenous legal productions of space are not the focus of this chapter.

Legislatively Producing 'Indigenous Spaces' in Post-Soviet Russia: the Why

Political and economic transformations ensuing from the end of the Soviet period have entailed an all-embracing re-conceptualization of Russia's spaces by the Russian state, as a necessary part of the processes in the state's transition toward marketization and democratization. A new property rights system was critical to economic growth and attracting foreign investment. Meanwhile, growing international concern over indigenous rights caused the Russian state to include a re-conceptualization of indigenous spaces as a small part of its extensive land and resource reform programme (Fondahl and Poelzer 2003, Xanthaki 2004, Fondahl 2005). Borrowing loosely from Andrew Woolford's (2005) cogent argument regarding the two impulses, *justice* and *certainty*, that currently drive the State in the Canadian province of British Columbia to re-define *where is indigenous*,[8] I argue that these impulses are at work as well in the Russian Federation's re-conceptualization of indigenous spaces within its borders, if in markedly

7 See, for example, von Benda-Beckmann, von Benda-Beckmann and Griffiths (2009), Blomley (1994, 2004, 2010) and Blomley, Delaney and Ford (2001).

8 Woolford's (2005) discussion of these impulses focuses specifically on the case of treaty negotiations in British Columbia, Canada.

distinct ways from that of Canada. Justice, in Woolford's argument, has to do with addressing issues of past disregard and denial of rights, including those of dispossessing indigenous peoples of their lands (2005: 3, 7–10). In Russia, progress on this 'justice front' has come from both internal pressure – the mobilization of indigenous intelligentsia and sympathetic non-indigenous scholars and legal practitioners in demanding legal protection for indigenous peoples – and from external initiatives to compel the State to observe the emerging international norms and principles regarding indigenous rights.

Woolford identifies the creation of new state–indigenous relations as currently pursued in British Columbia as focusing on creating certainty about a path into the future that promises a heightened well-being for all citizens (2005: 3, 10–12). *Certainty* is lacking where tenure is not clear – where ownership over land and resources is not well-defined. *Certainty* that does not provide well-being for one party is likely to be short-lived, as the dispossessed party will probably at some point revolt, causing a return to uncertainty.

The Russian Federation has pursued a policy of re-visualizing its total space concordant with new market relations and private property interests. Within this much larger project, it has re-visualized indigenous space. I argue that the impulse of *justice* was more manifest in the first decade of post-Soviet legislative developments regarding indigenous spatialities, and that the first decade's legislation also focused on *certainty* for indigenous peoples and livelihoods, in terms of economic well-being – even if through a rather narrow filter of the so-called traditional forms of land use. In the second decade we see increased emphasis on *certainty* for investors: the state seeks to encourage capital investment in the Russian North and reduce capital flight from this area.

Moreover, the Russian Federation, in the spaces it has conceptualized, has failed to accommodate the very quintessence of indigenous peoples' territorialities, a point I will return to briefly at the end of this chapter.

The First Decade

Where is indigenous? Where should indigenous peoples be (given specific territorial rights)? Where should they be (spatially) empowered?

Shortly before the demise of the Soviet Union, the Russian Soviet Federation of Socialist Republics (RSFSR) adopted a Land Code (Land Code 1991), which guaranteed indigenous peoples rights to use their traditional lands, and to practice their so-called traditional economic activities. This led to the development of a variety of legal acts addressing the spaces perceived necessary to these activities (hunting, reindeer herding, trapping, fishing, marine-mammal hunting, gathering). It can be argued that Russian law in the first decade visualized two main types of spaces for indigenous peoples, both intimately linked to these indigenous activities.

Firstly, Russian law envisioned the creation of discrete, protected territories on which groups of indigenous persons (families, extended families, larger collectives) organized into *obshchinas* (roughly translating into communities or communes) could carry out their traditional activities. These bounded, discrete territories would remain the property of the state, but would be allocated to the *obshchina* for use and management 'in perpetuity' and without charge (rent), and protected from industrial development (Edict 'On Urgent Measures' 1992, Federal Law 'On ... Obshchinas' 2000). Limited self-governance rights over these territories were recognized.

A second form of indigenous territoriality visualized by law was called the Territory of Traditional Nature Use (TTP, using the abbreviation of *territoriia traditsionnogo prirodopol'zovaniia*). Such territories were to be set aside to protect more extensive tracts of land for traditional indigenous activities, protected against industrial encroachment – a sort of ecosystem-based approach to safeguarding the territories needed for such activities. The TTPs might include *obshchina* lands (Edict 'On Urgent Measures' 1992, Federal Law 'On Territories of Traditional Nature Use' 2001).

The Russian Federation's legislation[9] thus has mostly equated indigenous spaces to spaces of so-called traditional activities. This elided with the definition of indigeneity adopted in the Russian Federation, in which one of the four criteria in the law guiding *who* is indigenous is the 'pursuit of a traditional way of life and economy'. In imagining *obshchinas* and TTPs as appropriate forms of indigenous territoriality, Russian law relegated indigenous peoples spatially to supposedly *natural* landscapes, particularly to forest and tundra. The texts of the laws also explicitly connected traditional activities with nature protection, thus producing *obshchinas* and TTPs as sites of environmental protection. Spaces needed for the perpetuation of these traditional activities were to be set apart from the spaces of a marketizing economy, protected from 'mainstream' economic developments.

During this same decade a number of federal laws on natural resources – forestry, subsurface resources, the continental shelf – also addressed indigenous spatial rights, usually in one or a few articles. The spaces of Russia's forests were made available to indigenous peoples in order that they might access the forest resources that served their 'traditional activities' and supported their 'traditional way of life', without charge (Forest Code 1997). That is, preferential use rights were granted to indigenous peoples, though the forest resources remained state property. Subsurface resources also remained the property of the state, but the impact of their exploitation on traditional activities had to be taken into consideration, and a part of the income from the resources extracted in areas inhabited by indigenous

9 Early legal developments in the Russian Federation regarding indigenous spaces have been summarized elsewhere in the English-language literature (Fondahl 1998, Fondahl et al. 2001, Fondahl and Poelzer 2003, Osherenko 2001, Xanthaki 2004). Texts of the legislation, along with a more in-depth analysis and discussion of the legal developments is available in Russian in Kriazhkov (1994, 1999 and 2005).

peoples had to be directed to supporting the socio-economic development of these peoples (Federal Law 'On Subsurface Resources' 1995). Plans for exploiting both organic and mineral resources of the continental shelf had to consider the interests of indigenous peoples, and indigenous peoples whose way of life or employment depended on the living resources were given priority rights to these resources (Federal Law 'On the Continental Shelf' 1995). Once again, the extensive nature and requirements of traditional indigenous activities and their potential sensitivity to the environmental degradation of even parts of the territories used are recognized in these federal laws.

One other type of space was legislated, though not exclusively for indigenous peoples. A law from the end of the Soviet period stipulated that local self-government could be established (Law 'On Local Self-Government' 1991). Such governments would enjoy powers over the territories attached to their jurisdictions. In areas where indigenous groups predominated, such self-governments could become, in essence, local *native* self-governments. While 'ethnic territories' for indigenous northerners had existed since the 1930s, when a number of autonomous *okrug*s were established,[10] by the end of the Soviet period the indigenous population of these areas constituted a minority, in some cases comprising well under ten per cent of the *okrug*'s population. Local 'ethnic' governments, including those for indigenous peoples, had been created at the village and *raion*[11] level in the 1920s, but these had been discontinued by the late 1930s. The new federal legislation (and corresponding sub-federal laws in some cases) gave rise to numerous local native self-governments at the village level, and a few at the *raion* level (for example, Fondahl 1998, Fondahl, Lazebnik and Poelzer 2000, Fondahl et al. 2001).

We can assess these laws of the Russian Federation's first decade as founded in part on impulses of *justice* – to bring Russia closer to complying with international indigenous rights legislation and norms (as called for in the Constitution) and to protect and support so-called traditional elements of culture among populations that had suffered assimilatory policies for the previous half-century. It should be underscored that the framework laws recognize that the protection, preservation and allocation of lands is necessary to protect traditional activities and the identities intimately connected to these activities. Protection and allocation does not stem from rights flowing to indigenous peoples due to their being the 'original' inhabitants on the territory, as it does in Canada's First Nations.

Of course imperatives for *certainty* were at work during this period as well: in the newly evolving economy that encouraged privatization and marketization it was important to define what was *native space* in order to identify what is *not native space*', and therefore open to other uses, including most notably natural resource development. It was also important, as noted above, to enhance indigenous self-sufficiency through the support of traditional activities, to decrease the need for

10 The Nenets, Khanty-Mansi, Yamal-Nenets, Taimyr (Dolgan-Nenets), Evenki, Koriak, and Chukchi Autonomous Okrugs. An *okrug* is similar to a province.

11 A *raion* is similar to a county or district.

subsidizing a sparse, widely distributed and remote population. In the Russian Federation indigenous communities experienced significant 'neotraditionalism' or 'demodernization' during this period as the state withdrew supports, and 'traditional activities' took on a greater and sometimes very critical importance for survival (Pika 1999).

The Second Decade

If during the 1990s, the Russian State through legal acts conceptualized new types of indigenous spaces such as *obshchinas* and TTPs, during the 2000s, it became evident that attempts to materialize some of these spaces – most notably TTPs – would meet with failure. Moreover, during this second decade the attributes of the indigenous spaces, as originally visualized in law, underwent legal revision (Kriazhkov 2010, Øverland 2009, Riekkinen 2011, Yakel and Khmeleva 2005). Vladimir Kriazhkov, a legal expert on indigenous rights in Russia and advisor to the Russian Federation's constitutional court, has referred to this period as one of 'legal stagnation' for indigenous rights (2012: 29); 'legal retrenchment' may be an even more appropriate term, as the new laws have curbed rights provided for in the earlier legislation. Increasingly, indigenous spaces were either reduced/undermined or fully effaced in Russian law. We might say that the answer to *where is indigenous?* in the Russian Federation, as far as federal legislation's discourses are concerned, started to look more and more like *nowhere*.

In 2001, a revised federal Land Code was passed which effectively removed numerous provisions from federal legislation that had earlier applied to indigenous peoples' legal re-territorialization (Land Code 2001). No longer were land allotments available to indigenous peoples for the pursuit of traditional activities 'in perpetuity', nor could they enjoy them 'without charge'. Provisions for the preferential (non-competitive) allocation of land allotments to indigenous peoples for hunting, fishing and other traditional activities disappeared from the new legislation. Space became, at the discursive level at least, less ethnically differentiated.[12]

Industrial projects planned for territories inhabited by indigenous peoples were no longer required to evaluate the negative impacts on the traditional activities of indigenous peoples. Moreover, the stipulation that a portion of the income from the exploitation of subsurface resources be directed to the socio-economic development of indigenous peoples was removed (Kriazhkov 2012: 30, see also Kriazhkov 2010: 288ff).

It is also worth noting that over the past decade, no federal TTPs have been created: all attempts to do so have failed. Kriazhkov attributes this directly to the

12 Interestingly, such provisions for use of land without charge were retained in the 2006 version of the Forest Code (Forest Code 2006). The new code, however, did appear to prioritize forest companies' rights over those of indigenous peoples (Riekkinen 2011).

lack of 'sub-legal, normative acts' – those subordinate legislative acts needed to progress from the visualization of spaces (as per framework laws) to administration and materialization (Kriazhkov 2012, see also Øverland 2009).

Another momentous change is the dissolution of a number of the *ethnic territories*' (autonomous *okrugs*) created during the Soviet period. These autonomous *okrugs* had recognized the indigenous peoples, and provided them some modicum of rights and representation and, indeed, territoriality (Kuoljok 1985: 70–85). While such power had been rather limited during the Soviet era, in the post-Soviet period, the *okrugs* were seen as potential sites of increased political power. Though indigenous peoples had comprised minorities in these *ethnic territories* in terms of population, they nevertheless, to some small degree, were able to pursue agendas protecting their interests in these same territories. The Law 'On the Guarantee of Rights of Indigenous Peoples' (1999) provided for representation quotas for indigenous peoples to be established at the regional level. Arrangements for the representation of indigenous peoples, in various manifestations, have been created in the Yamalo-Nenets and Khanty-Mansi Autonomous Okrugs, and in the Sakha Republic (Yakutia) (Kriazhkov 2010: 344ff).

However, in 2007, the Taimyr (Dolgano-Nenets) and Evenki Autonomous Okrugs were subsumed by the Krasnoiarsk Krai, and the Koriak Autonomous Okrug was merged with Kamchatka Oblast' into a new unit – Kamchatka Krai. In losing such *ethnic territories*, the eponymous Dolgan, Nenets, Evenki, and Koriak peoples suffered forfeitures on several fronts. The *ethnic territories* had their own legal and governmental structures, enjoyed direct representation in the Russian Federation parliament, and negotiated directly with a host of federal agencies (Kriazhkov 2012: 30). While representing indigenous peoples' interests was not always the primary concern of the governments of these *ethnic territories*, the new, non-ethnic structures are far less likely to support indigenous interests.

Conclusion

Twenty years on, *where is indigenous* in the Russian North remains very uncertain. Vladimir Kriazhkov summarizes the situation: 'The existing legislation on indigenous numerically small peoples, if it is considered qualitatively, is declarative, contradictory, and deficient' (2012: 30). The lack of subordinate normative laws, at the federal and regional levels, that map implementation details has resulted in the materialization of such spaces being regionally uneven and indeed negligible in many areas. Moreover, federal legislation on indigenous rights to land and resources has been characterized by retrenchment in provisions regarding indigenous opportunities.More importantly, Russian law has failed to conceptualize indigenous spatialities that are indeed applicable to indigenous lived experiences and spatial practices. A large part of the indigenous population is urban, and is no longer involved in traditional activities. Yet even for those who remain 'on the land' the conceptualization, administration and materialization of discrete, bounded territories, such as *obshchinas*, ignore age-old adaptive strategies of movement across space to deal with changing resource distributions due to weather, climate, epizootics, etc. These conceptualizations ignore that over the course of seasons or years the same geographic space may be used by more than one group or people.

Tenure over discrete spaces is only one dimension of indigenous spatiality. Brian Thom, discussing the Coast Salish of British Columbia, talks of trajectories determined by both potential resources elsewhere, and the social networks that govern access to these resources. 'Individuals experienced their territories as "itineraries" of places engaging in reciprocal practices relating to their use, and respect, of the land within an ecosystem that they continually appropriate through their lives' (Thom 2009: 186). Anderson (2000), Sirina (2006, 2011) and Stammler (2005), to name just a few examples, illustrate indigenous spatialities in various parts of Siberia similar to those Thom describes for southern British Columbia. In legislatively designating spatial categories of *where is indigenous*, we see the Russian state discursively producing more liberal spaces in its first decade, and constricting these spaces in its second decade. Of course law is only an initial step in materializing such spaces. To date, *obshchinas* have been created in some parts of the Russian North (though many have ceased to exist after a short period); federal-level TTPs have yet to be established. However, critically, the spaces discursively provided for by law fall short of providing for indigenous spatialities. *Where is indigenous?* to be answered justly and with certainty for both indigenous and non-indigenous citizens of Russia, needs to engage much more intensively with the complexities and specificities of indigenous spatialities.

Acknowledgements

I warmly thank Joachim Otto Habeck for his kind invitation to present an earlier version of this paper at the conference on *Nomadic and Indigenous Spaces: Productions and Cognitions* – and in doing so, for luring me back, perhaps a bit faster, from university administration to the professoriate. Thanks are also due to Vladimir Alekseevich Kriazhkov who over the years has been very generous in providing me with copies of his articles and books on Russian law regarding indigenous peoples.

References

Anderson, D. 2000. *Identity and Ecology in Arctic Siberia: The Number One Reindeer Brigade*. Oxford: Oxford University Press.

Blomley, N. 1994. *Law, Space and the Geographies of Power*. New York: Guilford.

Blomley, N. 2004. The boundaries of property: Lessons from Beatrix Potter. *Canadian Geographer*, 48(2), 91–100.

Blomley, N. 2010. Cuts, flows and the geographies of properties. *Law, Culture and the Humanities*, 7(2), 203–16.

Blomley, N., Delaney, D. and Ford, R. 2001. *The Legal Geographies Reader*. Malden, MA: Blackwell.

Donahoe, B. 2006. Who owns the taiga? Inclusive vs. exclusive senses of property among the Tozhu and Tofa of southern Siberia. *Sibirica*, 5(1), 87–116.

Donahoe, B., Habeck, J.O., Halemba, A. and Sántha, I. 2008. Size and place in the construction of indigeneity in the Russian Federation. *Current Anthropology*, 49(6), 993–1009.

Edict 'On Urgent Measures'. 1992. Ukaz Prezidenta Rossiiskoi Federatsii 'O neotlozhnykh merakh po zashchite mest prozhivaniia i khoziaistvennoi deiatel'nosti malochislennykh narodov Severa' ot 22 aprelia 1992 g., N°397 [Edict of the President of the Russian Federation 'On urgent measures for the defence of the places of habitation and economic activity of the numerically small peoples of the North', 22 April 1992, no. 397]. *Vedomosti s"ezda narodnykh deputatov Rossiiskoi Federatsii i Verkhovnogo soveta Rossiiskoi Federatsii*, 18, 1009. (Reproduced in part in Kriazhkov 1994: 199).

Federal Law 'On the Continental Shelf'. 1995. Federal'nyi zakon 'O kontinental'nom shel'fe Rossiiskoi Federatsii' ot 30 noiabria 1995 g., N°187-F3 [Federal Law 'On the Continental Shelf of the Russian Federation', 30 November 1995, no. 187-F3]. *Sobranie zakonodatel'stva Rossiiskoi Federatsii*, 49, 4694. (Reproduced in part in Kriazhkov 1999: 95).

Federal Law 'On ... Obshchinas'. 2005. Federal'nyi zakon 'Ob obshchikh printsipakh organizatsii obshchin korennykh malochislennykh narodov Severa, Sibiri i Dal'nego Vostoka Rossiiskoi Federatsii' ot 20 iiulia 2000 g., N°104-F3 [Federal Law 'On the General Principles of the Organization of Obshchinas of Indigenous Numerically Small Peoples of the North, Siberia and the Far East of the Russian Federation', 20 July 2000, no. 104-F3]. *Sobranie zakonodatel'stva Rossiiskoi Federatsii*, 30, 3122. (Reproduced in Kriazhkov 2005: 59–72).

Federal Law 'On the Guarantee of the Rights'. 1999. Federal'nyi zakon 'O garantiiakh prav korennykh malochislennykh narodov Rossiiskoi Federatsii' ot 30 aprelia 1999 g., N°82-F3 [Federal Law 'On the guarantee of the rights of indigenous numerically small peoples of the Russian Federation', 30 April 1999, no. 82-F3]. *Sobranie zakonodatel'stva Rossiiskoi Federatsii*, 18, 2208. (Reproduced in Kriazhkov 2005: 48–56).

Federal Law 'On Subsurface Resources'. 1995. Federal'nyi zakon 'O nedrakh', Zakon Rossiiskoi Federatsii v redaktsii Federal'nogo zakona 'O vnesenii izmenenii i dopolnenii v Zakon Rossiiskoi Federatsii "O nedrakh"' ot 3 marta 1995 g., N°27-F3 [Federal Law 'On subsurface resources', Law of the Russian Federation in the version of the Federal Law 'On changes and additions to the Law of the Russian Federation "On subsurface resources"' of 3 March 1995, no. 27-F3]. *Sobranie zakonodatel'stva Rossiiskoi Federatsii*, 10, 823. (Reproduced in part in Kriazhkov 1999: 92).

Federal Law 'On Territories of Traditional Nature Use'. 2001. Federal'nyi zakon 'O territoriiakh traditsionnogo prirodopol'zovaniia korennykh malochislennykh narodov Severa, Sibiri i Dal'nego Vostoka Rossiiskoi Federatsii' ot 7 maia 2001 g., N°49-F3 [Federal Law 'On territories of traditional nature use of indigenous numerically small peoples of the North, Siberia and the Far East of the Russian Federation', 7 May 2001, no. 49-F3]. *Sobranie zakonodatel'stva Rossiiskoi Federatsii*, 20, 1972. (Reproduced in Kriazhkov 2005: 72–7).

Fondahl, G. 1998. *Gaining Ground? Evenki, Land and Reform in Southeastern Siberia*. Boston: Allyn & Bacon.

Fondahl, G. 2005. 'Everything is as if beyond a boundary'. Reflections on apprehensions regarding aboriginal re-territorialization in Northern Russia, in *Rebuilding Identities: Pathways to Reform in Post-Soviet Siberia*, edited by E. Kasten. Berlin: Dietrich Reimer, 89–107.

Fondahl, G., Lazebnik, O. and Poelzer, G. 2000. Aboriginal territorial rights and the sovereignty of the Sakha Republic. *Post-Soviet Geography and Economics*, 41(6), 401–17.

Fondahl, G., Lazebnik, O., Poelzer, G. and Robbek, V. 2001. Native 'land claims', Russian Style. *Canadian Geographer*, 45(4), 545–61.

Fondahl, G. and Poelzer, G. 2003. Aboriginal land rights in Russia at the beginning of the twenty-first century. *Polar Record*, 39(209), 111–22.

Forest Code. 1997. Lesnoi kodeks Rossiiskoi Federatsii ot 29 ianvaria 1997 g., N° 22-F3 [Forest Code of the Russian Federation, 29 January 1997, no. 22-F3]. *Sobranie zakonodatel'stva Rossiiskoi Federatsii*, 5, 610. (Reproduced in part in Kriazhkov 1999: 112).

Forest Code. 2006. Lesnoi kodeks Rossiiskoi Federatsii ot 4 dekabria 2006 g., N°200-F3 [Forest Code of the Russian Federation, 4 December 2006, no. 200-F3]. *Sobranie zakonodatel'stva Rossiiskoi Federatsii*, 50, 5278.

Habeck, J.O. 2005. *What it Means to Be a Herdsman: The Practice and Image of Reindeer Husbandry among the Komi of Northern Russia*. Münster: LIT.

Kedar, A. 2006. On the legal geography of ethnocratic settler states: Notes towards a research agenda, in *Law and Geography*, edited by J. Holder and C. Harrison. Oxford: Oxford University, 401–41.

Kriazhkov, V.A. 1994. *Status malochislennykh narodov Rossii: Pravovye akty i dokumenty* [Status of the numerically small peoples of Russia: Legal acts and documents]. Moskva: Iuridicheskaia literatura.

Kriazhkov, V.A. 1999. *Status malochislennykh narodov Rossii: Pravovye akty* [Status of the numerically small peoples of Russia: Legal acts]. Moscow: Iuridicheskaia literatura.

Kriazhkov, V.A. 2005. *Status malochislennykh narodov Rossii: Pravovye akty. Kniga tret'ia* [Status of the numerically small peoples of Russia: Legal acts. Third volume]. Moskva: Iuridicheskaia literatura.

Kriazhkov, V.A. 2010. *Korennye malochislennye narody Severa v rossiiskom prave* [The indigenous numerically small peoples of the North in Russian law]. Moskva: Norma.

Kriazhkov, V.A. 2012. Rossiiskoe zakonodatel'stvo o severnykh narodakh i pravoprimenitel'naia praktika: Sostoianie i perspektivy [Russian legislation on Northern peoples and the practice of its application: Current state and prospects]. *Gosudarstvo i pravo*, 5, 27–35.

Kuoljok, K.E. 1985. *The Revolution in the North: Soviet Ethnography and Nationality Policy*. Uppsala: Acta Universitatis Upsaliensis.

Land Code. 1991. *Zemel'nyi kodeks Rossiiskoi Sovetskoi Federativnoi Sotsialisticheskoi Respubliki*, priniat 25 aprelia 1991 g. [Land Code of the Russian Soviet Federative Socialist Republic, passed on 25 April 1991]. (Relevant sections reproduced in Kriazhkov 1994: 122–5).

Land Code. 2001. Zemel'nyi kodeks Rossiiskoi Federatsii ot 25 aprelia 2001 g., N°136-F3 [Land Code of the Russian Federation, 25 April 2001, no. 136-F3]. *Sobranie zakonodatel'stva Rossiiskoi Federatsii*, 44, 4147. (Reproduced in part in Kriazhkov 2005: 78–81).

Law 'On Local Self-Government'. 1991 [1992]. *Zakon Rossiiskoi Federatsii 'O mestnom samoupravlenii v Rossiiskoi Federatsii'* ot 6 iiulia 1991 g., N°1550-1, tekst s izmeneniiami i dopolneniiami 1992 g. [Law of the Russian Federation 'On Local Self-Government in the Russian Federation', 6 July 1991, no. 1550-1, with changes and additions as of 1992]. (Reproduced in part in Kriazhkov 1994: 126–8).

Lefebvre, H. 1991. *The Production of Space*. Oxford: Basil Blackwell.

Martin, D.G., Scherr, A.W. and City, C. 2010. Making law, making place: Lawyers and the production of space. *Progress in Human Geography*, 34(2), 175–92.

Morris, P. 1999. *Negotiating the production of space in Tl'azt'en territory, 1969–1984*. Masters Thesis, University of Northern British Columbia.

Morris, P. and Fondahl, G. 2002. Negotiating the production of space in Tl'azt'en territory, northern British Columbia. *The Canadian Geographer*, 46(2), 108–25.

Osherenko, G. 2001. Indigenous rights in Russia: Is title to land essential for cultural survival? *Georgetown International Environmental Law Review*, 13, 695–734.

Øverland, I. 2009. Indigenous rights in the Russian North, in *Russia and the North*, edited by E. Wilson Rowe. Ottawa: University of Ottawa, 165–85.

Pika, A., ed. 1999. *Neotraditionalism in the Russian North: Indigenous Peoples and the Legacy of Perestroika*, edited in English by B. Grant, with a new afterword by B. Prokhorov. Edmonton: Canadian Circumpolar Institute.

Riekkinen, M. 2011. Participatory rights of Russia's indigenous peoples regarding land issues. *Issues of Business and Law*, 3, 110–20.

Sirina, A. 2005. Clan communities among the northern indigenous peoples of the Sakha (Yakutia) Republic: A step to self-determination?, in *Rebuilding Identities: Pathways to Reform in Post-Soviet Siberia*, edited by E. Kasten. Berlin: Dietrich Reimer, 197–216.

Sirina, A. 2006. *Katanga Evenkis in the 20th Century and the Ordering of their Life World*. Edmonton: Canadian Circumpolar Institute.

Sirina, A. 2011. *Problemy tipologii i preemstvennosti etnicheskikh kultur evenkov i evenov (konets XIX – nachalo XXI vekov)*. Aftoreferat dissertatii [Problems of typology and succession of ethnic cultures of the Evenkis and Evens (late 19th and early 20th century). Dissertation summary].
Slezkine, Y. 1994. *Arctic Mirrors: Russia and the Small Peoples of the North*. Ithaca, NY: Cornell University Press.
Sokolovski, S. 2003. The 'indigenous peoples' category in Russian legislation. *Anthropology News*, 44, 6–7.
Stammler, F. 2005. *Reindeer Nomads Meet the Market: Culture, Property and Globalization at the 'End of the Land'*. Münster: LIT.
Thom, B. 2009. The paradox of boundaries in Coast Salish territories. *Cultural Geographies*, 16(2), 179–205.
Von Benda-Beckmann, F., von Benda-Beckmann, K., and Griffiths, A. 2009. *Spatializing Law: An Anthropology of Law in Society*. Farnham: Ashgate.
Woolford, A. 2005. *Between Justice and Certainty: Treaty Making in British Columbia*. Vancouver: University of British Columbia Press.
Xanthaki, A. 2004. Indigenous rights in the Russian Federation: The case of the numerically small peoples of the Russian North, Siberia and Far East. *Human Rights Quarterly*, 26, 74–105.
Yakel, Y. and Khmeleva, E. 2005. Legal defense of indigenous peoples' rights to establish areas of traditional nature use. *ANSIPRA Bulletin*, 13/13a, 32–4.

Chapter 5
The Nellim Forest Conflict in Finnish Lapland: Between State Forest Mapping and Local Forest Living

Nuccio Mazzullo

The Sámi people are the European Union's only indigenous people, and they inhabit Northern Fennoscandia. Their relationship to the land has been shaped by their reindeer herding livelihood. In the past, reindeer herding was characterized by a long migration, which required a deep knowledge of the landscape. This traditional land use has undergone major changes and today often competes with other forms of land use. This chapter will discuss conflicts arising from large-scale forestry in the reindeer herding area of Nellim in the Northeast of Finnish Lapland. It will present an ethnographic case study[1] of a conflict that mainly revolved around the access to resources in that territory, which has been defined as state forest and is being managed by the Finnish Forest and Parks Service, Metsähallitus.

This chapter reflects on several aspects of the conflict that developed over decades, and in particular the role of maps in the negotiations between Metsähallitus and Nellim reindeer herders. It will be argued that maps are being used instrumentally to define spaces for certain uses, and exclude others (Lefebvre 1991). The abstract space, projecting future uses, represented in the maps by Metsähallitus will be juxtaposed with the sketches of actual land use by reindeer herders, representing lived experience. The two representations, one abstract, one lived experience, however, are not equally equipped with (scientific) authority, thus putting reindeer herders into a weak position. At a point in time when reindeer

---

1 The fieldwork for this study was conducted in the years 2009 and 2010 in the Nellim area as part of the project *Power Technologies' Productions of Space: Sámi Territoriality and Indigeneity* at the Leibniz Institute for Regional Geography (IfL) in Leipzig, Germany, in collaboration with the Arctic Centre, University of Lapland, in Rovaniemi, Finland. The project is part of the Collaborative Research Centre (Sonderforschungsbereich) 586 *Difference and Integration*, funded by the German Research Foundation (DFG). The fieldwork was, in its final part, conducted in conjunction with my current research project *Oral History of Empires by Elders in the Arctic* (ORHELIA), at the Arctic Centre, funded by the Finnish Academy. This chapter presents excerpts of interviews in Finnish and Sámi, which have been translated by the author. After each quote, the year and the running number of the interview file are indicated as well as the location of the quote in the transcript in terms of line numbers.

herders realized that they could not achieve their goals, they turned the discussion towards issues of indigeneity, which helped them to gain a stronger stance in the negotiations with the state authority. Complicating the picture is the ambiguous role of Metsähallitus, which has commercial interests yet simultaneously is supposed to act as the forest protecting institution. As a state enterprise, it is also obliged to protect indigenous culture, as defined in the Finnish constitutional law.

Sámi Reindeer Herding in Finland

The Sámi people inhabit the northern regions of the three Nordic states of Norway, Sweden and Finland as well as the Kola Peninsula in Russia. Their total population is estimated to consist of about 100,000 people. In Finland, their number ranges between 7,000 and 9,000, of which about 2,000 speak a Sámi language. Today, only 40 per cent of them live in the Finnish Sámi homeland, *Sápmi* in the Sámi language. The main centres are Inari, Enontekiö and Utsjoki, the three northernmost municipalities of Finland, forming the region of Finnish Upper Lapland (Raento and Husso 2001).

Reindeer herding along with hunting and gathering has always played a central role as both economic and cultural activities and, in different degrees, has remained a traditional livelihood for the majority of the Sámi. As such, reindeer herding is still recognized as a central and fundamental component of Sámi culture. However, it has lost some of its economic importance and it is practised in conjunction with other seasonal activities such as fishing, trapping, hunting and nowadays also tourism (Pietikäinen and Kelly-Holmes 2011) as well as other state and private employment.

Before the closure of national state borders in the nineteenth century, preventing the herders to cross them with their reindeer herds, Sámi had practised reindeer herding over large territories that stretched across the Norwegian, Swedish and Finnish borders (Aikio, Aikio-Puoskari and Helander 1994: 37–8, see Koch, this volume, Koch and Miggelbrink 2011). The negotiation of Finnish national borders entailed dramatic changes in the lives and society of reindeer herders, affecting social institutions like the *siida*[2] kinship patterns and alliances. Restricted mobility drastically reduced the semi-nomadic life of the herders and enforced a shorter seasonal migration between summer and winter pastures (Mazzullo 2010: 104).

2 The term *siida* is a Sámi term which refers simultaneously to a herding band, its herd and its pasture. Originally, the *siida* was kinship-based, but has undergone major changes with the introduction of the *paliskunta* (Mazzullo 2010: 110–12). Lehtola defines *siida* as a 'community of family groups who share and govern a jointly owned territory' (2002: 88). It 'comprises a small number of families who reside and migrate together, and who co-operate in the management of an aggregate herd of individually owned stock' (Ingold 1978: 150). According to Helander's definition, the notion of place is important, as the term *siida* also refers to 'a camp or herd belonging to one or several persons or families; dwelling place and home, place where the family tents are' (Helander 1999: 13).

In 1898, joint pasture cooperative systems in the form of *paliskunta* were introduced to reindeer herding, representing the beginning of a centralized and rigorous administrative control. Most importantly, the organization of reindeer herding in *paliskunta* clashes with the previous practice of free pasture herding within kinship-based bands, as highlighted by Ingold (1980). In addition, the creation of national parks, as well as the protection of predators such as wolves, bears and wolverines, has exerted significant pressure on reindeer herding (Müller-Wille et al. 2006: 367).

Following the subordination of Sámi reindeer herding under the Ministry of Agriculture and Forestry with the first Reindeer Herding Act in 1932, reindeer herding had to pursue the same market logic as agricultural activities. Furthermore, Sámi reindeer herders are facing increasing pressure to adopt new national and European regulations on how to practise herding and to make their practices economically viable notably because of the high costs of technological support (snowmobiles, helicopters, global positioning system devices), and even more so in bad winters when supplementary feeding becomes necessary. Many Sámi reindeer herders criticize that the current legislation for Sámi reindeer herding obliges them to find a mode of production that is more coherent with, and favourable to the Finnish type of reindeer *farming* as practised in the southern areas of the Finnish reindeer herding area (Müller-Wille et al. 2006: 35). Still, statistically, the importance of herding is diminished as it is weighed against large-scale production of crops or cattle.

In contrast to Norway and Sweden, where Sámi people have exclusive rights to practise herding, most reindeer owners in Finland are Finnish farmers who herd reindeer as a secondary economic activity (Mazzullo 2010: 108). Also within the Finnish Sámi region (Map 5.1), most of the reindeer herders are ethnic Finns. Of the 5,000 reindeer herders in this region only 25 per cent are Sámi.

In its traditional form, 'Sami reindeer herding is founded on direct physical benefits derived from hides, antlers, meat, heart, liver, blood, etc.' (Müller-Wille et al. 2006: 38). Until the beginning of the twentieth century the Sámi were almost self-sufficient and able to barter for other products (salt, flour, sugar, coffee and fabrics). Thus, the introduction of the market economy including money transactions had a significant impact on reindeer herders' livelihoods. However, according to many herders, fundamental changes in herding techniques as well as the integration of a series of modern technologies, such as fencing, snowmobiles, telephones and GPS, have not eradicated the particular relationship to the reindeer. The relationship remains central to Sámi culture as it entails a particular understanding of the environment that does not distinguish between *nature* and *culture*, a notion based on Western traditions (Ingold 2000, Mazzullo 2010).

Forestry in Finnish Upper Lapland

Forest growth is very slow at the latitude of Finnish Upper Lapland, and the turnover – the time for a forest to regrow – can take up to 80 years. Old-growth forest offers the best conditions for reindeer pasture, which requires undisturbed, light areas for the lichens to grow.[3] Nevertheless, forestry has been very extensive at times, especially in the 1930s, when the region of Petsamo still belonged to Finland and offered access to an ice-free harbour in the Arctic Ocean. After the Second World War, the forest in Lapland was intensively exploited for post-war reconstruction and war damage repayment to the Soviet Union (Myllyntaus 1991). During those times, forest harvesting relied on horses. The Sámi people participated in forestry work as reindeer were used in areas inaccessible to horses.

Only the rationalization of forestry practices after the 1960s and especially in the 1980s implied a qualitative change in terms of damage to reindeer pastures. Furthermore, with the introduction of the forest harvester – a machine that measures, cuts and transports the trees – Sámi forest workers and their horses or reindeer were no longer needed, as a former Sámi forest worker explained. He had been working for Metsähallitus for over a decade, but he quit when their activities changed dramatically and, in his opinion, became 'mindless':

> Back then, I had the first motor saw. We were in the forest, where we were cutting trees in the *hevossavotta*[4] style. And then, suddenly, forestry changed in such a way that it was not worth towing the single trees over the mountain. ... We started to bring down the forest, hectare after hectare. The trees were knocked down, and we were not supposed to touch them at all. We were not supposed to take care of where they would fall. Big areas of good and healthy forest were brought down. (Interview 2010:0196, lines 702–10)

The new forestry style implied that Sámi herders no longer had a say in harvest practices, which in turn meant that they could no longer try to minimize the deleterious impact harvesting had on the pasture. Considering the negative impact

3 Lichens are most important in the winter season. In the harshest time of winter, when the ground is frozen and reindeer cannot dig for the ground lichens (*Cladina spp.*), they turn to lichen species that hang down from tree branches, such as the horsetail lichen (*Alectoria spp.*). When forests are felled, this emergency food is lost. In addition, if the ground is not well frozen the forest harvester machines destroy the ground lichens. If the ground is ploughed then there is basically no vegetation left. The area needs more than a decade to recover before it can be used as reindeer pasture.

4 The term *savotta* describes a former forestry livelihood. Forest workers would bring their own horses, together build a hut in the forest and work continuously in an area for a few months. Their work on site included the choosing, cutting, cleaning and transportation of the single trees to a specific location from where the trees would then be taken to factories. *Hevossavotta* refers to the forest livelihood practised with horses (*hevoset*), in contrast to *savotta* practised with reindeer in areas where horses were useless.

commercial forestry has on reindeer pastures, the region of Lapland has experienced conflicts between herders and Metsähallitus since the end of the Second World War. However, these conflicts never gained media attention until the 1980s when environmental organizations campaigned for protecting old-growth forests in the northernmost (and most sensitive) parts of the boreal forest. The Kessi forest, a forest near the village of Nellim, was suggested to be turned into a national park. A wider conflict over rights to land developed in the following, involving civil society at large (Lehtinen 1991: 81–3), and has never been resolved.

Nellim Village

The village of Nellim is situated in the municipality of Inari, on the easternmost shore of Lake Inari about eight kilometres from the border with Russia. The territory of the Inari municipality is the largest of the three northernmost municipalities of Finland. About 90 per cent of Lapland's territory consists of state-owned forest. The economy of this region is characterized by traditional reindeer herding together with some hunting and gathering activities as well as forestry. In addition, many residents are employed by the municipalities or the state.

Nellim is a village of about 200 residents who speak one Sámi language (Inari Sámi or Skolt Sámi) and Finnish. The Skolt Sámi community came to Nellim after the Second World War as part of the plan to resettle those who were evacuated from the Petsamo area (see Map 6.3 in Koch, this volume). The Skolt Sámi had experienced resettlement before: with the Peace Treaty of Tartu in 1920, they had been relocated into the then Finnish territory of Petsamo. Then, with the ceding of Petsamo to the Soviet Union they were offered to remain Finnish citizens. Those coming from Petsamo were resettled into the villages of Nellim and Keväjärvi, and those coming from Suenjel village were resettled into the newly built village of Sevettijärvi north of Lake Inari (Ingold 1976). The people who were relocated to Nellim actually returned to some of the traditional Skolt Sámi hunting grounds southeast of Lake Inari. Thus, their relation to the region has direct importance for their official status as indigenous people, as originally this region was their homeland (Lehtola 1997: 66).

In the inter-war period (1920–39), well before Skolt Sámi were to be resettled in the village, Nellim had seen a period of relative growth in connection to the construction of the Arctic Ocean road. However, with the loss of the Petsamo region in the 1950s it witnessed a slow decline. Notwithstanding the influx of Skolt Sámi people, the community has shrunk considerably. While over a hundred pupils who attended the local school in that decade, nowadays there is only one pupil who is taken by taxi to school in the nearest town of Ivalo. The old village school of Nellim has now been turned into a hotel and wilderness centre to serve tourists seeking the *true wild northern nature* and the exoticism of the only indigenous people of Europe in organized adventure safaris.

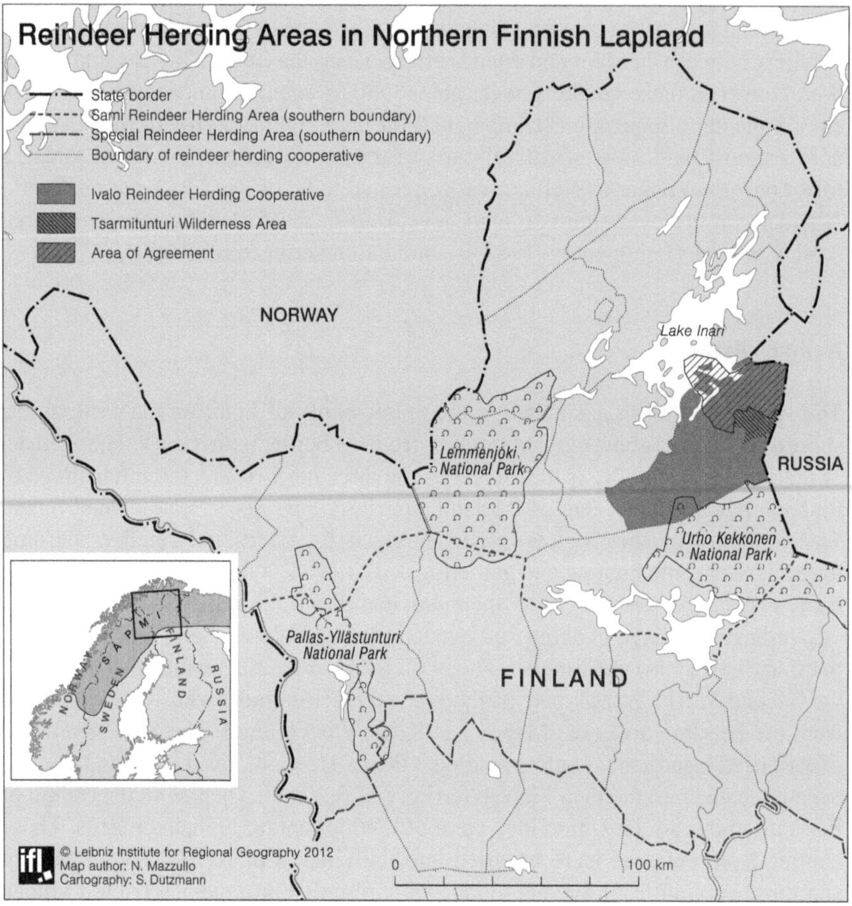

Map 5.1 Reindeer herding areas in northern Finnish Lapland

The Nellim Conflict

Until today, the forest around the village of Nellim is 'one of the northernmost localities in the world where commercial forestry is practiced' (Raitio 2008: 5). Being home to Sámi reindeer herders, conflicts over land use have affected the community for decades and troubled residents of Nellim especially since 2001.

In 2001, the simmering conflict had been revived with an open letter by the Kemi-Sompio herding cooperative south-west of Nellim addressed to the Ministry of Agriculture and Forestry. In the letter, members of the cooperative complained about current forestry methods and their negative impacts on pastures and thus, reindeer herding. The following year, a delegation of four herding cooperatives from the municipality of Inari, including the Nellim herding-group, visited the Ministry of Agriculture and Forestry in Helsinki to submit a document that highlighted the negative impacts of forest harvesting practices on northern pastures. The document also requested ways to minimize the damages and outlined areas where harvesting could continue, although less intensively to preserve precious old-growth forest and with it, a herding livelihood (Saarela 2003: 8, see also Greenpeace 2005: 7, Raitio 2008: 173).

In 2002, the Ministry of Agriculture and Forestry appointed an official to arbitrate the dispute between reindeer herders and Metsähallitus. Pirkko Saarela, who is currently a manager at the Lapland Regional Centre for Economic Development, Transport and the Environment (ELY-keskus), made an effort to mediate between the two parties. The following year, a report was published making suggestions on how to resolve the conflict (Saarela 2003). These suggestions, for example, to implement a moratorium on cuts in some areas regarded particularly important, were brought forth to the attention of the Ministry of Agriculture and Forestry. The so-called *Saarela process*, however, failed to achieve any immediate, tangible results as the ministry did not follow her advice.

In mid-December of 2003, Metsähallitus continued with the planned harvesting in the forests of Kirakka and Kessi. However, soon after, the Ministry of Agriculture and Forestry prepared a response plan to the Saarela report. The response was to deepen the research and to revise Metsähallitus's Natural Resources Plan (Metsähallitus 1999) with the aim of taking the proposed protection plans into account (Raitio 2008: 166). At the end of December 2003 new meetings were held where Metsähallitus signalled that it wanted to consider the issues proposed in the document and report, including the possibility of a moratorium in those crucial pasture forests.

In her report, Saarela had described the situation in Nellim quite dramatically, sketching out how twenty years of logging had seriously damaged most of the winter pastures forcing reindeer to leave their habitual pastures and to try to go south in search of forage (Saarela 2003: 32). Despite the urgency described in this report, no progress was made until the beginning of 2005, when negotiations were resumed with the unchanged aim of reindeer herders to halt the felling in the most important grazing grounds within the cooperatives' territories (Raitio 2008).

In February 2005, the World Wildlife Fund for Nature (WWF), the Finnish Association for Nature Conservation and Greenpeace published a letter of support in favour of the herders' demands to stop felling in the important grazing areas. Metsähallitus responded to that letter that felling was conducted according to the current Natural Resource Plan, which included consultation on the plans. However, reindeer herders claimed that there had been no real consultation; rather, they had just been informed of the plan.

This event marks the beginning of a phase in the conflict in which Greenpeace became involved as an active player. In particular, the conflict escalated dramatically after Greenpeace set up a monitoring station, called the 'Forest Rescue Station'. Activists started to mark the conflict area with banner posts which read 'reindeer forest'. In response to this action, the group of forest workers who felt that their jobs were threatened by this campaign set up a counter-camp that bore the banner of 'Anti-Terror Info Center'. The confrontation between the two camps – shown in a documentary, *Last Yoik in Saami Forests* (directed by Hannu Hyvönen, 2007), that was later released – continued until about three weeks later when Greenpeace removed the station to continue the campaign in other ways.

In June 2005, during negotiations between Metsähallitus and the representatives of the herding cooperatives, it became clear that in fact Metsähallitus did not have a mandate that would have allowed them to make an agreement. The meeting stopped with the question: how was it possible that the Ministry of Agriculture and Forestry had argued since 2002 that the conflict should be resolved at the local level, whereas when the meeting finally took place it was only to discover that the competences to settle the conflict remained at the ministerial level (Raitio 2008)?

The felling continued despite demands to stop until the conflict had been settled. At this point, in October 2005, three reindeer herders (all brothers) from Nellim took the conflict to a new level and requested the District Court of Lapland to issue a moratorium on the felling in their pasture. The court issued the moratorium, but at the same time asked the Sámi herders to pay a compensation for the loss of income incurred by Metsähallitus. While Metsähallitus tried to determine the amount of compensation that it would incur it continued its forest activities. The herders then filed a complaint to the UN Human Rights Committee about the continuing forest cuts despite a moratorium. This added a new political dimension to the conflict as a month later, in November 2005, the Committee warned the Finnish government to stop the felling in the Nellim area,[5] albeit to no avail. In 2006, the Finnish government carried on to support Metsähallitus's continuation of the activities in Nellim as, from the government's point of view, no human rights had been infringed. At this point, backed by two moratoria, the three Nellim Sámi herders filed legal action against Metsähallitus to the District Court of Lapland (Raitio 2008: 183).

In the meantime, Greenpeace's campaign was having an effect on the market as industrial customers buying paper products from conflict areas found themselves worrying about their image. In a press release, the Sámi Council informed that the Italian Banca Etica had ticked off Stora Enso, the paper-producing company in which the Finnish state is a majority shareholder, from the list of ethical investments.[6]

In July 2008, the court decided on the case put forward by the three Nellim Sámi herders. Their demands were rejected and they were sentenced to pay Metsähallitus a compensation of 99,000 euros, a verdict they appealed.

5 See press release by the Sámi Parliament (16 November 2005): Human Rights Committee order Finland to stop logging activities in Sámi area. Available at: http://www.saamicouncil.net/?newsid=1816&deptid=1169&languageid=7&news=1 [accessed 12 November 2012].

6 Press release by the Sámi Council (8 June 2007): Italian Ethical Fund excludes Stora Enso because of Inari wood. Available at: http://www.saamicouncil.net/includes/file_download.asp?deptid=&fileid=2865&file=Stora%20Enso%20PRESS_RELEASE_2007_06_08.pdf&pdf=1 [accessed 12 November 2012], see also Miggelbrink and Mazzullo (2011).

In early 2009, Metsähallitus announced its intention to implement forestry activities in the herding cooperatives of Muotkatunturi and Paatsjoki, despite the complaint of local herders. However, these plans did not materialize as in August 2009 the actors came to an agreement that was heralded as a victory by all involved parties.The end of the Nellim conflict was reached on 20 August 2009, when finally the actors agreed to sign a contract that sufficed as a compromise. It determines that for about 16,000 hectares of the 27,000 hectares of the conflict area a moratorium will be enforced for the period of 20 years. In the remaining 11,000 hectares soft felling methods shall be practised. As part of the deal the Nellim Sámi brothers do not have to pay the 99,000 euros in compensation to Metsähallitus to which they had been sentenced. From all sides – the herders, Greenpeace, Metsähallitus, Stora Enso, the Union of forest owners, etc. – this agreement is perceived to give enough time to jointly find better or more durable solutions without jeopardizing the importance of those pastures and forests that will be spared by the forestry activities in the meantime (Miggelbrink and Mazzullo 2011).

The Different Roles of Metsähallitus

The conflict described above is not an isolated one, as there have been plenty between Sámi people and national governments over the exploitation of the forest resources in the past and today all over *Sápmi* (Lehtinen 1991, Magga 2003, Raitio 2006). Given that the state land in the North amounts to almost 90 per cent of the whole region the state can decide where and when to fell a forest that has been reported by state experts to be mature for harvest.

The complaints made by reindeer herders did not dismiss the exploitation of the forest regarding the harvest as such, but criticized the extensive damages produced during the felling process, affecting the forest low vegetation and lichens, and even more afterwards due to the deep ploughing of the ground. More specifically, in the Nellim conflict, the main contestation not only concerns the use of the resources, but also the decision-making power on economically overlapping areas. These areas are defined as winter pastures by the Nellim herding group, whilst conversely understood and then mapped as state forests for the Ministry of Agriculture and Forestry, and as commercial forest area for Metsähallitus. Both forestry and herding are regulated by the Ministry of Agriculture and Forestry. Within this institutional setting, Metsähallitus represents the ministry's executive hand. At the same time, Metsähallitus is also in charge of issues regarding the protection of the forest which are regulated by the Ministry of the Environment.

Hence, Metsähallitus has to pursue two parallel and contradicting perspectives, carrying two different hats, so to speak. When Metsähallitus wears the environmental-agency hat, biodiversity and untouched landscapes are the aim and the protection policies are the means to obtain those aims, possibly after mapping its territory. When it wears the state-enterprise hat, the forest is its primary resource, and forest felling is the way to make profits to satisfy its shareholder demands. The ambivalent role of Metsähallitus surfaces in negotiations and when, in the public arena, its aims and views must continuously be re-adjusted to suit the situation and the institutional interlocutors.

During the period of the Nellim conflict, lasting about two decades, no concrete actions were taken towards the resolution of the conflict on behalf of Metsähallitus. Herders had been complaining that in several places Metsähallitus was cutting down forest areas that were relevant to reindeer herding whereas Metsähallitus, in contrast, kept repeating that it was following a plan that met strictly scientific standards, assuring the herders that it was aware of the felling's impact on the environment and livelihoods. In contrast, Sámi herders were unable to make claims on the ground of a similar level of scientific *evidence*. Their views and plans are usually seen and understood by outsiders as mere improvisations and make-shift strategies in tune with the very nature of nomadic herding patterns.

The former president of the Sámi parliament, Pekka Aikio, sees a fundamental contradiction between the requirements of the state and Metsähallitus's practices, which put reindeer herders into a difficult position. In an interview in 2010, he told me that in one occasion, he

> ... proposed to Metsähallitus that if the government is telling us that we have too little pasturages for the amount of reindeer than we have now, it means that government is saying that we cannot give one square meter away. ... And it is not possible that you [Metsähallitus] ... you take one tree away from us because the government itself has said right now that we are on the edge. So, Metsähallitus, we cannot let you then take one single tree away. (Interview 2010:0214, lines 174–79)

Participation in the planning of forest cuts is almost non-existent. On one occasion, one of my informants experienced that employees of Metsähallitus looked at the map and thus determined the forestry plan for a certain period. Although obliged to negotiate plans with other groups of interest, Metsähallitus would only later announce the plans. In the case my informant referred to, specific forest areas in the territory of the Inari Municipality were announced to be felled. Following the public announcement of the plan, Metsähallitus would then go ahead, as it was the case in the Nellim forest, and cut a substantial amount of forest in a small area. Considering the invasiveness of the felling, it had a significant impact on the conditions of the pasture. Metsähallitus, however, argued that the cuts had not been exceptionally big considering the size of the municipality, and thus, the state institution blatantly denied that their activity had any kind of impact on the pastures. It was only when the small group of herders joined forces with Greenpeace that the course of events changed quite drastically. The reasons for such change are twofold:

- Greenpeace knows how to map a territory, and how to use a map as a tool to present the effects of different local economic activities to the wider (international) public;

- Greenpeace campaigned in the conflict between a state enterprise and an indigenous group so effectively that it changed customers' attitudes towards products from the area, which in turn forced international companies trading the products to react, thereby exerting an influence on the local situation.

During the process of negotiations the most relevant tool was the map. In the beginning, the only available maps were those by Metsähallitus, showing either protected areas or forest to be used economically. These maps were devoid of any mention of reindeer pastures in the vast areas of northern Lapland. By looking at them, it had to be assumed that reindeer would graze in any area, and hence, logging in a relatively small area would not affect grazing in general. However, the quality of pastures clearly depends on certain features of the landscape (and is affected by changes in this sensitive ecosystem), thus, cuts in important pastures have a significant effect on herding. When finally, with the help of Greenpeace, reindeer herders started to produce maps showing important grazing grounds, it was quite easy to see the economic interests of reindeer herders and those of Metsähallitus clashing in certain areas. And this started to change the course of events.

Figure 5.1 Pekka Aikio shows the critical pasture areas in the different reindeer herding cooperatives

Source: Photograph by Jacinthe Briand-Racine

Figure 5.2 Unfolding plans onto the map. Greenpeace activist Jarmo Pyykkö describing the Nellim conflict during an interview

Source: Photograph by Nuccio Mazzullo

More specifically, on the maps, Metsähallitus could show that most of the territory in Upper Lapland is protected (ironically, from its own activities as it has this double role of preserving and harvesting forest). However, what was not depicted in those maps is that most of these areas consist of swamps and mires, which means that they are not good for late-spring herding – the time of the year when the reindeer often experience difficulties finding food and hinge solely on horsetail lichen because the ground can be frozen solid and the hooves are not strong enough to break the ice. Only with the help of Greenpeace were the herders able to demonstrate that Metsähallitus was cutting down precious forest essential for the reindeer's survival. In addition, Greenpeace helped to raise international awareness of the ongoing cuts as a violation of human rights, thus rendering the local conflict over forest cuts an issue of Sáminess and indigeneity.

The Importance of Indigeneity, Sáminess and Place-Making

As a state authority, Metsähallitus was expected to respect the rights of Sámi reindeer herders. By not highlighting important pastures in its maps for forest cuts in the National Resources Plan for Upper Lapland (Metsähallitus 1999), however, the state authority undermined the basis of reindeer herding. Consequently, the state was threatening a traditional Sámi livelihood, herders claimed, thus undermining the integrity of their culture, which had been guaranteed to the Sámi by the Finnish constitutional law (The Constitution of Finland 731/1999, §17 and §121). Their objections had no repercussions among state officials.

In the following, by emphasizing the notion of indigeneity, Sámi herders were able to reach international institutions. They thus by-passed local, regional and national institutions, gained visibility and reached the aforementioned agreement on a moratorium on forest felling in some of their pastures in 2009. Sámi herders used the notions of territory and indigeneity in accordance with the definitions provided in the UN Declaration on the Rights of Indigenous Peoples and in the ILO *Convention 169 Concerning Indigenous and Tribal Peoples in Independent Countries*. The conflict over land use in Upper Lapland gained another layer of complexity when identity issues were raised. Having been turned into an ethnic issue, the local problem acquired a different dimension, and Metsähallitus suddenly had to deal with human rights claims. The level of contestation increased accordingly, also among different groups of the Finnish society as the following pages will show in reference to interview material.

In interviews I conducted during my fieldwork, it becomes quite clear that the conflict started to be ethnically driven after the interruption in 2002, when it was realized that Metsähallitus was not in charge of settling the conflict. The active involvement of Greenpeace and other activist groups brought the confrontation to the international level. It was this change of arena that forest workers perceived as the starting point of the conflict as Greenpeace was able to connect the conflict to already existing campaigns. In fact, the role that Greenpeace had was to make the voices of reindeer herders heard on an international level by claiming that indigenous peoples were being suppressed and that the Finnish government was violating human rights. The lobbying of environmental groups forced the Finnish government to deal with the local problem and to take a stand internationally on the moratorium, which in turn put pressure on Metsähallitus to act more carefully. As a Sámi reindeer herder interviewee put it in 2010:

> At the peak of the conflict, it was Greenpeace and all the others, so that they [Metsähallitus] no longer dared to continue as they used. They asked nicely enough, can we still cut and if so, where? And they went to cut there where we had told, though sometimes they went a little bit into unauthorized areas. So in fact, they would cut a little more. Well, they always wanted to get some more ... (Interview 2010:0113–14, lines 1440–5)

Another interviewee, who is a Finnish forest worker, did not show much sympathy with the activists, but he admitted that the arrival of Greenpeace had an impact on the latent conflict, 'let's say that maybe it speeded up the resolution a little' (Interview 2010:0199, line 958). The involvement of green activists sparked a debate about indigeneity among groups of Finnish society, illustrating that Sáminess is understood differently in different situations. When asked what the Nellim conflict was really about, a Finnish resident of Nellim replied:

> I think it was a matter of principle. I thought it was just the ILO ratification. Who does this land belong to? Who gets to decide? I think it was like that. Or rather it would be like a step in that direction. Finland has not yet ratified the ILO Convention. It drags on, yeah. And they do not dare, there are too many of these political controversies. Yeah, so, I think it was precisely the question of who is in charge here. There is nobody to whom these lands belong. Do we fell all the forests here and there? What is the forest, and what is nature, what are nature's values? Are they profits derived from cellulose? ... Surely it makes no sense, like, hundred year old trees are just being cut when you know what the forest means to people. The implications it has for the reindeer ... And here is just Kalevi, [one of the reindeer herders who filed a complaint against Metsähallitus; Nuccio Mazzullo] he's done well. For many years he has done research, for example, just like scientific research, asking what that means: Will a reindeer go to this or that place? And how is the lichen doing? And so on. Horsetail lichen plays an important role. But above all, it is a question of principle. (Interview 2010:0163, lines 538–59)

The interviewee continues to discuss the importance of Sámi reindeer herding for humankind:

> Well, yes, it seems to me like it is just the fact that they can practice the inherently natural livelihood that was part of their activities ... Now, reindeer husbandry is awfully tight, nowadays, it is just really tight ... there are a lot of contradictions the young generation has to face because of all these cooperatives' conflicts, and then you have these forests conflicts. Not many young people think that this is worth it anymore. I think it's really just a big damage to the whole Sámi culture if reindeer husbandry is lost. I think it must be thought of as a damage to the nation, to the whole world if indigenous cultures are lost. (Interview 2010:0158, lines 537–91)

Others critically raised questions of power when ethnicity was used strategically in forest conflicts. Asked whether there is any relation between ethnicity and herding and whether it had characterized the controversy, a Finnish interviewee argued that

> Yes indeed, I believe that these reindeer herders in general have gained huge power. For some reason, this livelihood, reindeer herding, the herders have gained so much ground. Precisely in this conflict, precisely these herders. Somehow, they have been given just so much power here. (Interview 2010:0186, lines 503–15)

However, the same informant responded to the question of whether Sáminess had anything to do with the conflict:

> I do not think that Sáminess has anything to do with this, … I think that everything started with reindeer herding, … because although they [Sámi] claim that what is traditional is not the way it is now. (Interview 2010:0186, lines 546–9)

An official from Metsähallitus, who himself is Sámi, highlighted how in the Ivalo herding cooperative more than half of the reindeer owners are Sámi 'and yet we had managed to reach an agreement, ... so there is not such a big difference' (Interview 2010:0123, lines 292–4). However, he agrees that in the Nellim case,

> it was fundamentally important that the applicants were Sámi reindeer herders, it is precisely the Sámi identity that was raised in court, and that is the most important thing. When we [Metsähallitus] cut the forest we harm the reindeer husbandry livelihood, and through that we harm Sámi culture, and hence, it becomes a human rights violation. Therefore, it is against the ILO Convention. With this action, they were successful in bringing Sáminess to the general attention. ... Sáminess, it's like a good tool. A tool that can be used against anyone. Not just against forestry, but also against the mining industry as well as tourism and anything that undermines Sámi culture. ... Sáminess is a good tool for a variety of disputes. It's like a tool. (Interview 2010:0123, lines 299–319)

In this quote, the Sámi interviewee recognizes that, on the one hand, Sáminess as such does not hinder the possibility of reaching agreements on issues regarding the planning of forest cuts in the cooperative's territory. On the other hand, he highlights that, given the current international legislation, Sáminess can be used as a discursive instrument (of course, only by indigenous reindeer herders, although non-indigenous herders may have been equally affected by the felling), and by that, concepts may acquire different weight and effectiveness in the legal resolution of conflicts.

In legal texts, the *official definition* of indigenous peoples reflects the place-boundedness of indigenous ways of living. Thus, the notion of (indigenous) identity is strongly connected to place. It gains its strength from the land and from the experience of places (Mazzullo and Ingold 2008). In this sense, the term *livelihood* describes engagement with the land, and is, as such, place-bound. Hence, reindeer herding could not be performed in other areas than the circumpolar North. As a matter of principle, as one of the interviewees put it (see above), only those people who had been living on and experienced that land could become reindeer herders. However, this knowledge remains a lived experience, and is not easily expressed in scientific terms, which makes it very difficult to be integrated into industrial forms of resource exploitation (Nadasdy 1999).

Maps and Space-Making

In the Nellim conflict, Metsähallitus's maps provided detailed accounts of forest composition for the sake of forest harvest. For an indigenous reindeer herder, these accounts had little relevance. However, backed with scientific information, Metsähallitus's maps retained a stronger position in the negotiations. For a long time, Metsähallitus could claim areas with the help of maps without being met by immediate strong contestation. Only when green activists contributed the necessary tools and networks, Sámi reindeer herders started to make claims both in form of maps and political pressure.

While the initial maps by Metsähallitus did not highlight areas crucial for reindeer herding, they did refer to place names used by reindeer herders to describe important features of the landscape. Place names are part of herders' intimate knowledge and hence belong to the lived experience of the place (Mazzullo and Ingold 2008). In local discourses in the North it is often pointed out that these lived experiences do not tell much to outsiders and even less to politicians and institutional actors in distant places, be it Helsinki, Brussels or New York. In order to make the location of crucial pastures visible during negotiations, reindeer herders sketched important areas or places on the maps provided by Metsähallitus. According to my informants, their information, however, would always remain a sketch. It would not achieve the same degree of validation that the printed, *official* information on the map had.

In my fieldwork during the years 2008–10, almost all conversations with informants would end up with a map being unfolded on a kitchen table. In the case of older Sámi herders, reading place names on the map would prompt them to tell narratives and to refer to lived experiences rather than to describe the borders of a territory, for instance of a protected area depicted on the map. For them, those names referred to a place that, in Ingold's words, 'is not a bounded portion of territory but a nexus of ongoing life activity' (Ingold 2005: 506). Lefebvre's notion of social space helps understand the complexities involved in the conflict evolving around cartographic representation in the case of the Nellim conflict.

According to Lefebvre (1991), space is not neutral, but fundamentally relational. Space is not a given, but a result of social interaction, hence, space is a social product. Practices define the mode of production, for instance, reindeer herding and timber production on forest land are two modes of production that define separate spaces. In the area of Nellim, the two spaces overlap. The overlapping spaces create conflict as the modes of production are incompatible. An integration of the two would require a new mode of production rather than the adoption of one or the other, because, Lefebvre argues, 'each mode of production has its own particular space, the shift from one mode to another must entail the production of a new space' (Lefebvre 1991: 46). Each new space is '"differential space", because ... a new space cannot be born (produced) unless it accentuates differences' (Lefebvre 1991: 52). The creation of a differential space in the Nellim area would mean that reindeer herding and forestry can find a common, new mode

Figure 5.3 A reindeer herder sketching migration patterns of the reindeer on a leisure activity map by Metsähallitus

Source: Photograph by Nuccio Mazzullo

of production that differs from each of the earlier modes. However, the description above gives evidence that the hegemonic group, understood in Gramscian terms, applied instrumental rationality in order to maintain their own mode of production. It engaged *abstract space* to eliminate existing differences, thus producing the impression of homogenous modes of production. For instance, on a map, both timber production forest and reindeer herding pastures may be subsumed under the term *commercial forest*, suggesting that a continuation of timber extraction does not exclude reindeer herding in the same area.

Thus, in the Nellim forest conflict we can observe the competition of different modes of production, finding their ground for confrontation in such abstract space. In the case of Nellim the map provided by Metsähallitus in the first place suggested future uses of the land's resources according to its mode of production. In the following, reindeer herders aimed at a representation of their lived experience by drawing sketches on those maps. However, only when they presented similar maps, their claims were taken into consideration.

Representing and recombining distinct social spaces, maps can become extremely important tools in the conflict over land ownership and uses. In general, each map has a purpose, a focus and an intended audience; thus, another map, one that depicts different properties, refers to a different social space. This is something that also indigenous peoples have become aware of and are increasingly making use of. As Pooles remarks,

> Maps have always been both symbols and instruments of power. After flag raising came the naming of places to express possession for the gratification of distant patron[s] of exploratory expeditions. Now, a revisionist tendency is reasserting itself: indigenous peoples are using maps to re-name and reclaim their lands. Their maps remain instruments of power, but a creative and restorative power. (Poole 1995: 1)

These maps are the result of institutionalized planning and printing and, according to Monmonier, '[a]t the root of their power is our frequently unquestioning acceptance of cartographic messages' (Monmonier 1995: 1). Sketches, on the contrary, are not maps. They are narratives of people's lived experiences in the landscape. When drawing the sketch, there is a relationship between the maker, the lines and the lived experience that is being re-enacted. However, in the Nellim case, the sketches of reindeer herders had never been officially validated, until they were transformed into a map that meets certain standards: printed, not sketched, with symbols and a legend, without handwriting. Thus, the sketches' validation existed only among those who were present at the time when they were made. Or, in the words of Denis Wood, '[a] sketch map, thus, is a map that fails to attain the level of presentation' (Wood 1993: 84).

The way the maps by Metsähallitus codified the Nellim forest entailed the recreation of the forest on a different level of abstraction, as 'a spatial code is not simply a means of reading or interpreting space: rather it is a means of living in that space, of understanding it, and of producing it' (Lefebvre 1991: 47–8). It was essential for the reindeer herders to submit their own map to contest the spatial code produced by Metsähallitus, thus achieving a moratorium on the felling of crucial pastures. Even though, this did not change the power relation between the parties involved.

At the same time, the experience the herders tried to communicate was actually a different one and cannot be conveyed through a map. Ingold (2000) analyses the differences between the lived experience of space through the accomplishment of different tasks and the abstract creation of space. In terms of the latter, the state achieves control through its various efforts to homogenize representations of landscapes, which are reduced to more simple forms, such as maps, making it easier to handle. In the description of the nomadic, indigenous attempt to characterize those spaces as places (Ingold 2000, 2011) we can see parallel arguments in the theoretical analytical framework of Lefebvre (1991), Ingold (2000), Deleuze and Guattari (2010) namely, the conceptual triad of the spatial practice, the representation of space and the representational space. According to Lefebvre, representation of space is 'conceptualized space, [it is] the space of scientists, planners, urbanists, technocratic subdividers and social engineers [who] identify what is lived and what is perceived with what is conceived' (Lefebvre 1991: 38). Metsähallitus occupies the conceptualized space that allows the production of maps. Finally, the representational space is 'space as directly lived through its associated images and symbols, and hence the space of "inhabitants" and "users"' (Lefebvre 1991: 38–9), such as the Sámi reindeer herders.

The contestation between Nellim herders and Metsähallitus corresponds to the difference between sketching and mapping: in the first case 'space is occupied without being counted', whereas in the second case 'space is counted in order to be occupied' (Deleuze and Guattari 2010: 18). 'Maps are a technology of power, and the key to internal power is cartographic process' (Harley 2001: 244). Maps serve the interests of those who make them.

Moreover, Wood defines maps as 'more or less permanent, more or less graphic artefacts that support the descriptive functions in human discourse that links territory to other things, advancing in this way the interests of those making (or controlling the making) of the maps' (Wood 2010: 20, see Wood and Fels 2008, Wood, this volume). At the same time, certain standards of map-making fundamentally restrict the possibilities of description.

By determining people's livelihoods in maps, a new space is produced and power exercised (Foucault 1984: 252). If we look at the maps produced by Metsähallitus, they are characterized precisely by this: an accurate description of the different types of livelihood that can be practiced in those areas and are connected by a network of forest roads used to reach those remote areas for resource extraction purposes. In other words, 'in the colonial project of occupation, this network spreads across the territory, overriding the tangled trails of inhabitants' (Ingold 2007: 81).

Concluding Remarks

From the material presented in this chapter, it is evident that the availability of maps determines the process of negotiating land uses. It has been argued that in the case of Nellim it was strategically important for indigenous peoples not only to rely on sketched presentations of their lived experiences, but to produce printed maps to match the format of representation with which they were confronted. Having authoritative maps of their own enabled them to access the 'abstract space' (Lefebvre 1991) in which Metsähallitus operated. Thus, the professional support of green activist groups to produce maps of reindeer herding pastures has been fundamental for Nellim reindeer herders to influence planning and decision making. The production of maps has not changed their relationship to the land, but it has equipped them with a thorough understanding of how to engage with officials, administrators and planners in an entangled system of state and semi-state institutions.

Maps were used as a strategic tool in the negotiations between Metsähallitus and reindeer herders. But only when the contestation over land uses had been redefined as an ethnic conflict using the idea of being an indigenous people, reindeer herders were able to use their maps effectively. Indigenous peoples are increasingly aware of the notion of indigeneity's potential strategic use to improve their stance in negotiations with state institutions. Those interviewees from Finland, quoted extensively above, share the understanding that such possibilities of empowerment are offered by the concept of Sáminess.

Acknowledgements

A special acknowledgement to the reviewers and to Hannah Strauss for the invaluable help and advices on the manuscript.

References

Aikio, S., Aikio-Puoskari, U., and Helander, J. 1994. *The Sámi Culture in Finland.* Helsinki: Lapin Sivistysseura.

Deleuze, G. and Guattari, F. 2010 [1986]. *Nomadology: The War Machine.* Seattle: Wormwood Distribution.

Foucault, M. 1984. Space, Knowledge and Power, in *The Foucault Reader*, edited by P. Rabinow. New York: Pantheon Books, 239–56.

Greenpeace 2005. *Lapland: State of Conflict. How the Finnish Government is Abusing the Forest Rights of Sámi Reindeer Herders* [Online]. Available at: http://www.greenpeace.org.uk/files/pdfs/migrated/MultimediaFiles/Live/FullReport/6889.pdf [accessed: 12 November 2012].

Harley, J.B. 2001. Deconstructing the map, in *Writing Worlds: Discourse, Text and Metaphor in the Representation of Landscape,* edited by T. Barnes and J. Duncan. London: Routledge, 231–47.

Helander, E. 1999. Sami subsistence activities – spatial aspects and structuration. *Acta Borealia: Nordic Journal of Circumpolar Societies*, 16(2), 7–25.

Ingold, T. 1976. *The Skolt Lapps Today*. Cambridge: Cambridge University Press.

Ingold, T. 1978. The transformation of the Siida. *Ethnos: Journal of Anthropology*, 43(3–4), 146–62.

Ingold, T. 1980. *Hunters, Pastoralists and Ranchers: Reindeer Economies and their Transformations*. Cambridge: Cambridge University Press.

Ingold, T. 2000. *The Perception of the Environment: Essays on Livelihood, Dwelling and Skill.* London: Routledge.

Ingold, T. 2005. Epilogue: Towards a politics of dwelling. *Conservation and Society* [Online], 3(2), 501–8. Available at: http://www.conservationandsociety.org/text.asp?2005/3/2/501/49324 [accessed: 12 November 2012].

Ingold, T . 2007. *A Brief Story of the Line.* London: Routledge.

Ingold, T. 2011. *Being Alive: Essays on Movement, Knowledge and Description.* London: Routledge.

Koch, P. and Miggelbrink, J. 2011. Being in the frontline of a Sámi Culture and a private business: Cross-border reindeer herding in northern Norway and Sweden. *Nomadic Peoples*, 15(1), 114–43.

Last Yoik in Saami Forests (dir. Hannu Hyvönen, 2007).
Lefebvre, H. 1991. *The Production of Space.* Oxford: Blackwell.
Lehtinen, A.A. 1991. Northern Natures – a study of the forest question emerging within the timber-line conflict in Finland. *Fennia*, 169(1), 57–169.
Lehtola, V.-P. 1997. *Saamelaiset – historia, yhteiskunta, taide* [The Sámi – history, society, art]. Aanaar-Inari: Kustannus-Puntsi.
Lehtola, V.-P. 2002. *The Sámi People: Traditions in Transition.* Aanaar-Inari: Kustannus-Puntsi.
Magga, H. 2003. Poronhoidon menetykset ja sopeuttaminen suuriin ympäristömuutoksiin Lapin paliskunnassa 1950-luvulta lähtien [Reindeer herding losses and adaptation to the biggest environmental changes in the Lapland Cooperative since the 1950s], in *Kuuluuko sääsken ääni taivaaseen?* [Does the sound of mosquitoes reach heaven?] edited by H. Heikkinen, Helsinki University of Technology, Laboratory of Environmental Protection, Publication 3/2003, 11–78.
Mazzullo, N. and Ingold, T. 2008. Being along: Place, time and movement among Sámi People, in *Mobility and Place: Enacting European Peripheries,* edited by J.O. Bærenholdt and B. Granås. Aldershot: Ashgate, 27–38.
Mazzullo, N. 2010. More than meat on the hoof? Social significance of reindeer among Finnish Saami in a rationalized pastoralist economy, in *Good to Eat, Good to Live with: Nomads and the Animals in Northern Eurasia and Africa,* edited by F. Stammler and H. Takakura. Sendai: Center for Northeast Asia Studies, Tohoku University, 101–19.
Metsähallitus 1999. *Natural Resource Plan for Upper Lapland.* Publications of Metsähallitus Forest Department, 38. Helsinki: Metsähallitus.
Miggelbrink, J. and Mazzullo, N.E. 2011. Winterweide und Holzlieferant: Indigene und andere Interessen im Konflikt über die Nutzung nordfinnischer Wälder [Winter pasture and timber provider: Indigenous and other interests in conflicts over utilization of northern Finnish forests]. *Geographische Rundschau*, 63(7–8), 36–42.
Monmonier, M.S. 1995. *Drawing the Line: Tales of Maps and Cartocontroversy.* New York: Henry Holt.
Müller-Wille, L., Hukkinen, J., Bölter, M., Müller, F. and Forbes, B.C. 2006. Synthesis: Environmental and socio-political conditions for modern reindeer management in Europe's North, in *Reindeer Management in Northernmost Europe: Linking Practical and Scientific Knowledge in Social-Ecological Systems*, edited by B.C. Forbes, M. Bölter, L. Müller Wille, J. Hukkinen, F. Müller, N. Gunslay and Y. Konstantinov. Berlin: Springer, 365–79.
Myllyntaus, T. 1991. *Electrification of Finland: The Transfer of a New Technology into a Late Industrializing Economy.* Worcester: Macmillan.
Nadasdy, P. 1999. The politics of TEK: Power and the 'integration' of knowledge. *Arctic Anthropology*, 36(1–2), 1–18.

Pietikäinen, S. and Kelly-Holmes, H. 2011. The local political economy of languages in a Sámi tourism destination: Authenticity and mobility in the labelling of souvenirs. *Journal of Sociolinguistics*, 15(3), 323–46.

Poole, P. 1995. Geomatics: Who needs it? *Cultural Survival Quarterly*, 18(4), 1–2.

Raento, P. and Husso, K. 2001. Cultural diversity in Finland. *Fennia*, 180(1–2), 151–64.

Raitio, K. 2008. *'You Can't Please Everyone'—Conflict Management Practices, Frames and Institutions in Finnish State Forests*. Ph.D. dissertation. Joensuu: University of Joensuu.

Saarela, P. 2003. *Selvitys Ylä-Lapin metsä- ja porotalouden yhteensovittamisesta: Työryhmämuistio* [Investigation on the coexistence of forestry and reindeer herding in Upper Lapland: Working group report]. Helsinki: Ministry of Agriculture and Forestry, 15.

Wood, D. 1993. What makes a map a map? *Cartographica*, 30(2–3), 81–6.

Wood, D. and Fels, J. 2008. *The Natures of Maps: Cartographic Constructions of the Natural World*. Chicago, IL: University of Chicago Press.

Wood, D. 2010. *Rethinking the Power of Maps*. New York: Guilford Press.

Chapter 6

Sámi–State Relations and its Impact on Reindeer Herding across the Norwegian-Swedish Border

Peter Koch

In the light of the volume's general focus on different understandings of space and conflicting forms of land use, this chapter analyses the roots of a conflict that has lasted for more than 250 years about reindeer pastures in the northernmost part of Europe. Indigenous pastoralists – Sámi[1] reindeer herders – have been migrating with their animals over a vast territory that came to be partitioned by four nation states in recent centuries (Norway, Sweden, Finland, Russia). The chapter's regional scope is limited to the Swedish-Norwegian facet of the multilateral debate around Sámi reindeer herding. What is peculiar here is that the territorial logic of the nation states (control over discrete and neatly bounded territorial entities) has been so pervasive that the Sámi themselves have got used to thinking and arguing in the state's terms, even though many of them continue to regularly cross the border with their herds. I explain the process of internalization of territorial logic with reference to Foucault's characterization of power relations. While his theories have already been applied by some authors (for example, Heikkilä 2006) to the situation of the Sámi in a fragmented political setting, the specific aim of this chapter is to apply Foucault's idea of three different power mechanisms – namely, juridical, disciplinary and security mechanisms – to analyse the history of the conflict. This permits us to discuss the conflicts about pastures, sovereignty and citizenship not so much as a sequence of historic events, but rather as a sequence and mixture of different logics of the state's attitude towards its subjects. I conclude with the statement that the third mechanism – the security mechanism – has thus far been applied only inconsistently, and that therefore the situation of cross-border reindeer herding remains precarious, as it is subject to fragmented and – as of yet – unresolved legal conditions.

Cross-border reindeer herding came into being in 1751 when the first border in the Sámi settlement area was established between Denmark-Norway and Sweden-Finland on the legal basis of the Lapp Codicil[2] (*Lappkodicillen*) attached to the

1 For an introduction to the Sámi see Mazzullo's chapter in this volume.

2 Full Swedish title: Första Bihang eller Codecill till Grense-Tractaten emellan Konungarikerne Sverige och Norrige, Lappmännerne beträffande.

border treaty. Therefore we should rather regard the border to *cross* traditional reindeer migration routes and pastures than vice versa. Traditionally, the migration cycles of the Sámi spanned a vast area with winter pastures in the forest areas of Finland and Sweden and summer pastures in the mountain and coastal areas of Norway.[3] Today, cross-border reindeer herding is only possible between Norway and Sweden on the legal basis of border conventions, while the borders with Finland and Russia were closed for reindeer migrations in 1852 and 1889 and have not been reopened since.

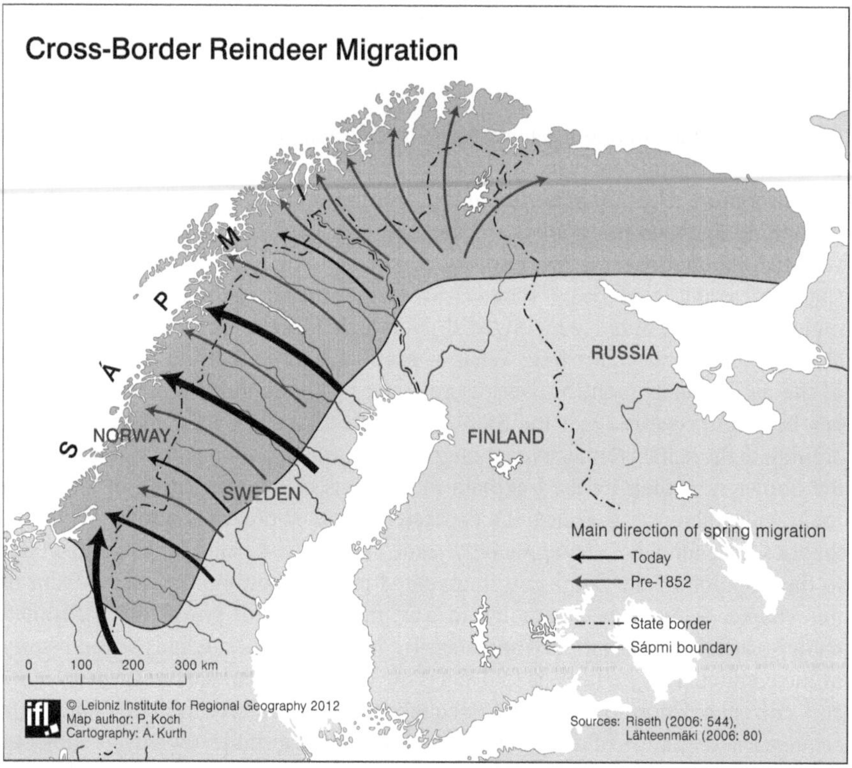

Map 6.1 Cross-border reindeer migration

Border conventions between Norway and Sweden were introduced in the twentieth century. They are supposed to run for a period of about 30 years, which corresponds with one generation. The last border convention from 1972 expired in 2005 but could not be replaced by a new convention. Instead Norway and Sweden created an unclear legal situation in which the older convention was unilaterally prolonged

3 For an overview of pastoral cycles in Fennoscandia see Ingold (1980), Koch and Miggelbrink (2011) and Storm (2009).

as a national law in Norway while in Sweden the Lapp Codicil was unilaterally reapplied. The consequences have been fatal: many Sámi from Norway could not access winter pastures in Sweden, while herders from Sweden used larger summer pasture areas in Norway and constructed new fences and corrals, which were then destroyed by the Norwegian police. The latter also pushed back the Swedish Sámi herders' animals with helicopters (Gruda 2007). Reindeer herders from Sweden demonstrated for their rights to land and water in Norway in Stockholm in 2007 and the governments found themselves confronted with severe international criticism.

How did this situation come about? Since negotiations over a new convention started in 1997, different national delegations composed by representatives from the Norwegian and Swedish governments and respective national Sámi institutions failed in achieving a mutually accepted consent on pastoral land rights across the state border. What appears to be a fundamental hindrance is the resistance of the local Sámi reindeer herding cooperatives on both sides of the border against an agreement that, from their point of view, puts their pastoral land rights and thus the basis of their livelihoods at risk. In an earlier article (Koch and Miggelbrink 2011), Judith Miggelbrink and I already addressed this conflict. We focussed on competing identifications and spatialities by analysing extensively empirical material from my fieldwork.[4] One of our most interesting findings is that the Sámi herders, irrespective of whether they live in Norway or Sweden, hardly ever employed any notion of Sámi indigeneity to express their economic interests and legal claims in this conflict. Totally in contrast, the Sámi herders acted and argued as citizens, Norwegian or Swedish, in the first place and in the second as Sámi inhabitants of the respective country, relying on the state to safeguard their particular business interests. This observation may be not so surprising if we consider the simple fact that the different citizenships divide the Sámi community, while the notion of indigeneity is common to the entire community and therefore did not serve as a decisive criterion in settling this conflict.

It seems obvious that nation-state citizenship has become a significant aspect of identity formation among Sámi individuals especially when it comes to pastoral rights across the Norwegian-Swedish border. I assume that each of the Nordic states actively worked toward such a situation in which 'the Saami were forced to adapt to dramatic legal, economic and political pressures' (Forrest 2002: 252) through policy making. The national states have created distinct legal reindeer husbandry regimes, intensive competitive reindeer industries, and thus citizenship identities also among the Sámi reindeer herders. I therefore believe it is necessary to again direct attention to Sámi–state relations, in particular to the states' roles to create state–related identifications among the Sámi. Many scholars have produced literature which predominantly presents historic overviews with a

4 Predominantly in-depth interviews with reindeer herders, Sámi servants and government officials that I conducted in Norway and Sweden from 2009 to 2011. For more information on fieldwork and the interviews see Koch and Miggelbrink (2011: 121–2).

chronological order of events. I shall instead reinterpret the rich literature with a different approach. I will order the events with a specific focus on cross-border reindeer herding with the help of Foucault's theory of power mechanisms. These mechanisms, I argue, should be regarded as guiding logics of state policies or, to put it in different words, as principles of how states deal with people and spaces. In my particular case it is about how states deal with an indigenous community and its homeland. In the following section I will first outline the characters of the three power mechanisms proposed by Foucault. I will then sort the events of Sámi–state relations according to each of the power mechanisms. In the conclusion I will return to Foucault and summarize my findings.

Power Mechanisms

Foucault (2007) first introduced the theory of power mechanisms in his lecture on *Security, Territory, Population* at the Collège de France in 1977/78. He denied that power was 'a substance, fluid, or something that derives from a particular source' and was therefore not in favour of 'a theory of what power is, but simply of power in terms of the set of mechanisms and procedures that have the role or function and theme, even when they are unsuccessful, of securing power' (Foucault 2007: 2). Power mechanisms, thus, are immanent to any social relation; and to analyse them, Foucault elaborates quite ambitiously, has the potential to 'initiate something like the overall analysis of a society' (Foucault 2007: 2).

Foucault identified three power mechanisms which are (1) the juridical, (2) the disciplinary and (3) the security mechanism (or apparatus of security) (Foucault 2007: 5–6). The *juridical mechanism* is concerned with the pure exercise of sovereignty over a populace within a territory that is characterized by a dichotomy of prohibition versus permission. It does not matter who those people are or what they want. They are not even *citizens* but simply *bodies* within the territory and must not violate the rules enforced in this territory; otherwise they will be ultimately punished. Sovereignty therefore represents 'a political relationship in that the decrees and laws must be implanted in the territory [in such a way] that no tiny corner of the realm escapes this general network of the sovereign's orders and laws' (Foucault 2007: 14). This is in line with how Forrest (2002: 251–2) characterizes a modern state's territoriality when he says '[t]he territorial model which generally typifies the modern state is a fixed (permanent), exclusive (belonging only to them), and geographically bounded (defined by hard lines) piece of land'. In my chapter, however, I shall show that prior to 1751–1826 the states in the North, although already *modern* at the time, lacked an exclusive territory but nonetheless already governed people, notably by taxing them, which led to a series of wars. So only the states' intention of gaining sovereignty over this *piece of land* necessitates the demarcation of an exclusive territory by creating borders, precisely because the juridical mechanism is already at play.

The *disciplinary mechanism* 'is a mode of individualization' (Foucault 2007: 12). People become individuals not just by being counted as single persons but also by being defined and classified as a person who is *someone*, that is, who belongs to a specific social group. Space accordingly becomes structured with spatial divisions that are designated for specific groups. The disciplinary mechanism 'belongs to the order of construction' (Foucault 2007: 17) and the (social) constructs are individuals and their spaces. Furthermore, the disciplinary mechanism is also about transformation, usually correction, of these constructs. It aims at 'eliminating its dangerous elements … and maximizing the good … by diminishing the bad' (Foucault 2007: 18) or simply at achieving perfection with all means.

The price for perfection may be too high, however. The *security mechanism* therefore operates with a calculation of costs aimed at achieving the 'optimal for a given social functioning' (Foucault 2007: 5). The optimal here is considered to be a statistically determined average which is accompanied by a tolerance range of the acceptable (Foucault 2007: 6). The goal is to maximize what is regarded positive and to minimize what is viewed as negative while acknowledging that perfection cannot be achieved. The identification of the optimal and the acceptable will require routines of monitoring, prevention, management etc. The security mechanism takes different variables into account that Foucault calls *uncertain givens*, notably the *population*, the *milieu*, and the contingency of *possible events*. Of the three variables I will pay particular attention to the milieu, defined by Foucault (2007: 21) as 'a set of natural givens – rivers, marshes, hills – and a set of artificial givens – an agglomeration of individuals, of houses, etcetera'. Foucault's term of the milieu goes far beyond the meaning of the term of the environment (which currently may attract greater attention in the social science community), for the milieu itself is not the given; rather, the milieu is produced purposefully with the givens to manage population through space.

According to Foucault (2007: 8), 'there is not a series of successive [power mechanisms], the appearance of the new causing the earlier ones to disappear. There is not the legal age, the disciplinary age, and then the age of security'. Instead within each mechanism one can find characteristics of the other. The appearance of the security mechanism, for example, will cause 'a real inflation of the juridico-legal code' and 'a considerable activation and propagation of the disciplinary corpus' (Foucault 2007: 7). Still, we have to consider that one and each mechanism is dominant during a given time, and the usual sequence is first the juridical, then the disciplinary and later the security mechanism.[5]

The power mechanisms – understood as logics of state policies toward people and spaces – can help illuminate the origins of this conflict if we recognize Sámi–state relations to be relations of power. In the following sections I will discuss how the Sámi, their lands and livelihoods came under state sovereignty (juridical

5 At least Foucault has not presented any example with an order that was vice versa in his entire work and I do not know of any other author who has done so.

mechanism), were identified as *Sámi* and transformed (that is, corrected and perfected) according to the needs of the states (disciplinary mechanism), and later acknowledged by the states as a political reality – to use Foucault's term, as an *uncertain* but *acceptable given* (security mechanism).

Juridical Mechanism: The Establishment of State Sovereignty in Northern Fennoscandia

In this section I will discuss how the states have implemented exclusive sovereignty in northern Fennoscandia by establishing state borders and integrating the Sámi population during the eighteenth and nineteenth century. In order to prepare the ground for this I will first briefly outline the situation when exclusive sovereignty and territories were still absent.

A Land of no Borders

The North Calotte has been home to various ethnic groups. As Hansen (2010: 1) specifies, 'a cultural diversification process among various heterogeneous groups of hunter-gatherers' took place, in which 'a separate Sami ethnic identity seems to have emerged during the last millennium B.C.'. At this time 'a separate *North Germanic* identity in the west' and 'the predecessors of the later Finnish population in the east' (Hansen 2010: 1) came into existence as well. Urbańczyk (1992: 164) mentions that 'it is generally accepted that stabilization of the nonegalitarian Germanic populations in the Arctic began during the Roman Age', that is, between 'ca. birth of Christ [and] ca. 400 AD' (Urbańczyk 1992: 164). The development of the Scandinavian states as political units gradually occurred in subsequent centuries. 'From 1328 to 1852', according to Sandberg (2008: 274), 'the geopolitical situation here was ambiguous and the national borders between Norway, Sweden, Finland and Russia somewhat unclear or not strictly enforced'. Indeed, from the Treaty of Nöteborg in 1323, we must consider the North Calotte rather to be a condominium (Tuchtenhagen 2009: 77) in which state territories, or taxation districts (Hansen 2010: 3) to be more precise, overlapped (Map 6.2). This situation allowed the Sámi reindeer herding communities to freely migrate between these territories but in many cases also subjected them to double or even triple taxation (Hansen 2005: 375, 377, Sandberg 2008: 274).

State Bordering and Territorialization

Fennoscandia underwent a long and gradual process of establishing state borders from the High Middle Ages to the nineteenth century[6] (Hansen 2005: 362).

6 Several border amendments took place between Russia and Finland in the Petsamo area in the course of Finnish independence and various wars in the first half of the twentieth

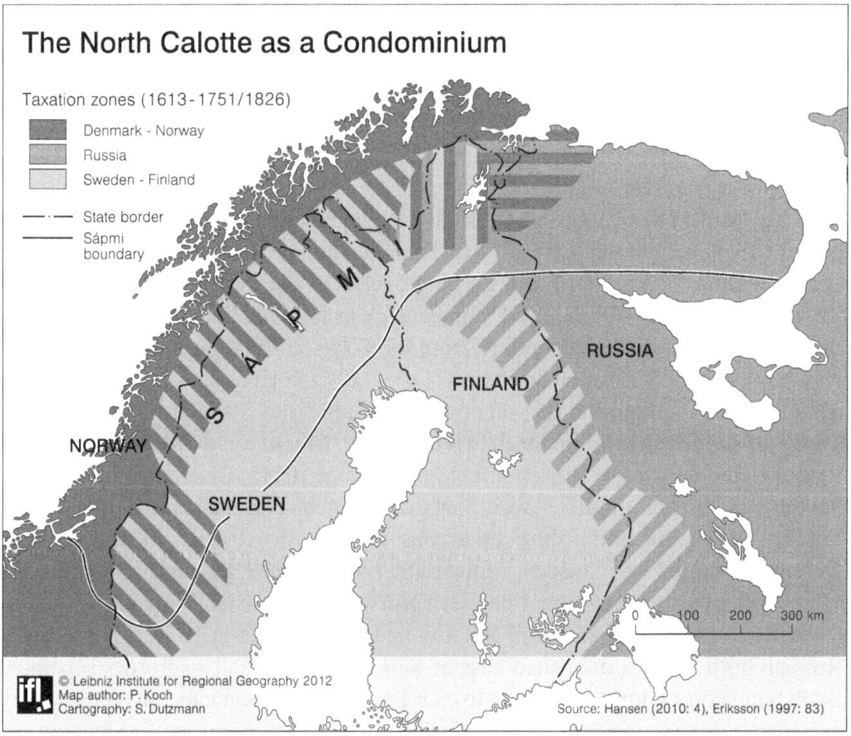

Map 6.2 The North Calotte as a condominium

Especially in the northern parts, the claims to power over areas by the three states of Denmark-Norway, Sweden-Finland and Russia widely overlapped. These states competed for control over the lands and its populations and resources. The absence of state borders and the unclear situation based thereon caused multiple partly military conflicts between the states throughout several centuries.

With the 1720 Treaty of Frederiksborg, which marked the end of the Great Northern War, Denmark-Norway and Sweden-Finland were finally determined to create a national border between the two countries from Nord-Trøndelag and Jämtland up north forever (Koch and Miggelbrink 2011: 116, Simowitsch 2010: 40, Storm 2009: 19). This area matches the traditional Sámi settlement area. Extensive research on the situation in what would become the borderland in the Sámi settlement area, known as the border examination minutes

century (see Mazzullo, this volume). We should also consider the various changes in the status of state borders in the North during the late twentieth century, for example, in the framework of the European Union (of which Sweden and Finland are members whereas Norway is not), the Nordic states' membership in the Schengen Agreement, and the European Union's Interreg and Northern Dimension policies.

(*Grenseeksaminasjonsprotokollene*), was conducted by Norwegian Major Peter Schnitler (Hansen 2005: 380, Prestbakmo 2007: 47–9, Storm 2009: 20) in 1742–45. These minutes formed the basis of the negotiations over the course of the first fixed borderline in this region, which resulted in the 1751 Treaty of Strömstad. The border was finally chosen to run along the mountain range and to incorporate all of Finnmark into Norway.

State borders with Russia had not been established until Finland, along with parts of Lapland situated east of river Torne, came under Russian rule as a Great Duchy within the Russian Empire in the course of the Finnish War or Swedish-Russian War in 1808–09. This war, however, was only the final episode in a series of Swedish-Russian armed conflicts from the 1700s to 1809, during which Finland was occupied by Russia several times. Originally the river Kalix was planned to demarcate the national border separating Finland from Sweden but in the 1809 Treaty of Fredrikshamn finally the rivers Torne, Muonio and Könkämä, situated further to the east, were chosen instead (Hedström 2011: 10).[7] As a consequence, not only Sámi communities were separated but also a substantial part of the Finnish population, today best known as Torne valley Finns (*Tornedalfinnar*), were incorporated into Sweden. At the same time, in 1809, the northern part of the 1751 border became the border between Norway and Russian Finland.

After a union of Sweden and Norway had been formed in 1814, a Norwegian-Russian border was established east of Kirkenes in 1826. This border separated Skolt Sámi communities and came to cause serious disagreement between Norway and Russia concerning the fishing rights of Finnish Sámi communities in summer. This disagreement resulted in the Russian borders being closed during the second half of the nineteenth century, which I will discuss later in this chapter.

States' Approaches of How to Deal with the Sámi Population and Reindeer Herding

With the establishment of the state border by Denmark-Norway and Sweden-Finland the governments 'recognised the existence of the transnational "Lappish nation"' (Eriksson 2002: 245) and its general rights to use ancestral lands and waters (Hansen 2005: 382), for which the so-called Lapp Codicil was attached to the 1751 Treaty of Strömstad (Koch and Miggelbrink 2011: 117, Mazzullo 2009: 177).[8] Thereby the two states acknowledged and affirmed the Sámi population's originally strong rights. Also compared internationally, 'the [Lapp] Codicil was

7 The parties agreed on the course of the border to follow the deepest points of the rivers, which, interestingly, requires a joint Swedish-Finnish commission every 25 years to verify and, as last happened in 2006, adjust the actual course of the border (Finnish Government 2007).

8 The Lapp Codicil is generally regarded as the Sámi Magna Carta (Forrest 2002: 257, Karppi 2001: 398, Koch and Miggelbrink 2011: 117, Marainen 1984: 9, Tuchtenhagen 2009: 76).

rather unique in defining the transborder territorial rights of a non-state group' (Karppi 2001: 399).

By and large, this meant the end of multiple and overlapping taxation of Sámi by the contracting parties, the states of Denmark-Norway and Sweden-Finland (Tuchtenhagen 2009: 78), which according to Marainen 'threatened the economic well-being of the Sami, and that the subsequent establishment of interstate borders therefore had positive effects for them' (cited in Eriksson 1997: 83–4). The exclusive taxation by one of the two states could be achieved by a provision in the Lapp Codicil over the implementation of state citizenship for the Sámi population. Individual Sámi groups could then choose whether they wanted to become either Norwegian or Swedish citizens and then to be subjected to taxation of the respective state.[9] The Lapp Codicil also granted the Sámi traditional rights to land and waters and full self-determination in all internal political, administrative and juridical matters and thus determined Sámi neutrality in case of war (Omma 2006: 43). However, what makes the Lapp Codicil the most important document concerning Sámi–state relations nowadays is the stipulation that the state border is not to hinder reindeer herding, entitling the Sámi to freely continue with their pastoral migration and the use of traditional pasture grounds with summer pastures in Norway and winter pastures in Sweden and Finland (Hansen 2005: 384, Tuchtenhagen 2009: 76).

What I think is most interesting about the Lapp Codicil is not the obvious fact that at this time the Scandinavian states came to recognize the Sámi, their traditional customs, their primary livelihood, reindeer herding, and their (counter) spatiality but its documentation of these states' general approach to dealing with the Sámi population. The overall goal of the states seems to have been to partition the Sámi population between the states but otherwise to leave them alone. This also seems to apply to the spatial dimension. As the state border essentially does not only cut through the entire traditional Sámi homeland in general but also through literally all pastoral migration routes of the mountain Sámi in particular, it seems that the states at this time were able to accept Sámi border crossings without any state intervention. One may regard this kind of border crossings as a permanent penetration of a state's border, which the states obviously could accept at this time. The states still lacked the means to enforce control in the north.

In contrast to the Lapp Codicil regulations, the Russian approach to establishing a border with Norway in the nineteenth century not only took into account but indeed encompassed the traditional migration patterns of the local Sámi. It thus created a situation in which regular border crossings in the framework of the regional reindeer pastoralism would not take place. Previously, the Russian government had made several attempts in vain to reach an agreement with the Norwegian

9 An interesting anecdote in this context is that in the beginning some Sámi successfully changed state citizenship seasonally (that is to be Swedish while staying on the territory of Norway and vice versa) and so completely avoided to be taxed by either of the two states (Marainen 1984: 7).

government that would allow the Finnish Sámi under Russian jurisdiction to fish in the Barents Sea in summer and build boathouses to store fishing boats in winter. It is said that the Norwegian parliament (*Stortinget*) and government mistrusted all Russian advances and had serious security concerns (Udtja Lasse 2007: 31, 39). Russia then even offered a sizeable portion of land northeast of Lake Inari to be incorporated into the Norwegian territory in 1859, which however was refused by both the Norwegian and the Swedish government (Udtja Lasse 2007: 35). As a consequence, Russia decided to fix the status of the Russian and Finnish borders with Norway, that is the complete closure from 1853.

This first border closing made migrations across the border with Russian Finland impossible for Norwegian Sámi. The Sámi of the Finnmark therefore lost access to their traditional winter pastures in the Finnish forests. In order to avoid these negative effects many Sámi declared themselves to be Swedish, which allowed them to continue with their traditional migrations from the winter areas in Finland to summer pastures in Norway via Sweden. This was possible since Russia had not yet closed the Finnish border with Sweden (Udtja Lasse 2007: 29). Yet with the ongoing disagreement between the Norwegian and Swedish governments and its Russian counterpart, the Finnish-Swedish border was also closed in 1889 which, already at this time, ultimately meant a stop of cross-border reindeer herding with Finland and Russia. Cross-border reindeer herding was henceforward only possible between Norway and Sweden. Pedersen thus argues that closing the border helped the Norwegian government to control Sámi reindeer herding (in Finnmark at least), which 'would ensure more favourable conditions for the Norwegian colonisation of Finnmark' (Pedersen 2002: 169).

These developments resulted in a massive relocation of Sámi groups, usually moving southward in Scandinavia, where border crossings were still and are until today possible, and also to the forests of Swedish and Finnish Lapland. These relocations caused disturbances to the ethnic and spatial structure of the Sámi (Map 6.3) especially in the more southern regions of the Sámi settlement area. North Sámi groups with their specific methods of reindeer herding – in particular the operation with larger herds – soon became very powerful within the new communities and – since the smaller herds started to follow the larger – came to cause a gradual though incomplete exodus of the original community members.

Disciplinary Mechanism: The Construction of Content to the States' Territories

In the previous section I discussed how the states in question implemented their sovereignty over the northern areas of Fennoscandia and how the Sámi and their livelihoods came under the respective state's jurisdiction. Mörkenstam (2002: 114), referring to Zygmunt Bauman, says 'the sovereignty of the modern state is, "the power to define and to make definitions stick – everything that self-defines or eludes the power-assisted definition is subversive"'. So, once the state is able to

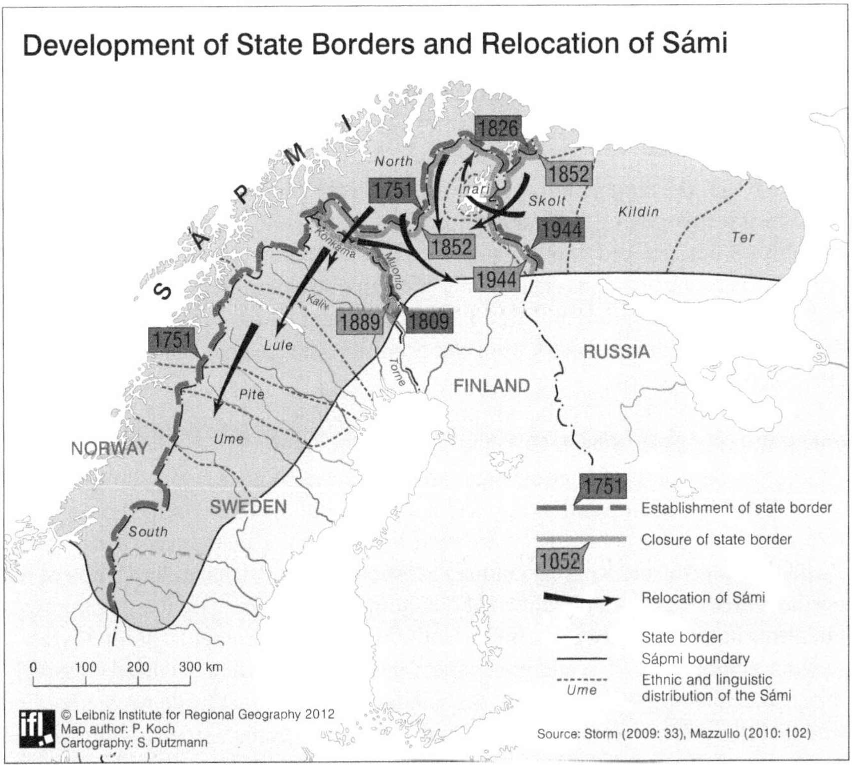

Map 6.3 Development of state borders and relocation of Sámi

install its sovereignty over areas and peoples, it has the power to define or identify its contents (in other words, to classify all what exists on its territory). Therefore I will now turn the focus on how Norway and Sweden literally put content into their newly fixed or demarcated territories so that this land actually became either Norway or Sweden and how the states transformed both the Sámi community into national branches and reindeer husbandry into part of the respective domestic economy. In a later step I will explain how the states finally became agents of *their* nationalized Sámi reindeer herding communities that then started to claim support from the respective state to advocate their interests as citizens in border convention negotiations.

Control over Territory: Colonization and Border Defence

As described above, the first border within the Sámi area or Sápmi was established between Norway and Sweden in 1751. However, colonization did not take place before the mid-nineteenth century and then soon threatened the indigenous community at large (Udtja Lasse 2007: 20) as the attitude towards the Sámi changed significantly based on the racist assumption that they represent an inferior culture. To me, a different source of state mistrust appears of even greater importance. The Sámi as a transnational group that had gained special rights through the Lapp Codicil such as free choice of citizenship, free migration over the state border with their reindeer herds and obligatory neutrality in case of war (which potentially hindered the development of identification with the respective state among Sámi in the first place) were assumed to not be reliable in protecting the newly established borders and newly gained territory. For this reason the governments attracted peasant Scandinavians to move to the northern areas by offering them temporary tax exemption and free land for farming as private property that had previously been dispossessed from the Sámi and transformed into crown land[10] (Pedersen 2006: 168).

Sandberg (2008: 274) calls this process of colonization in Scandinavia '"inner colonisation". This meant that the dominating agricultural and commercial cultures in the southern metropolitan areas of these countries spread northwards and pushed the hunting, gathering, and nomadic cultures of the indigenous groups into more marginal areas, or assimilated them into the mainstream Norwegian, Swedish, Finnish, and Russian cultures'. It should be noted that colonization of the northern areas was not only aimed at defending the periphery by the governments but also – and eventually more importantly from the governments' point of view – to reduce emigration of Scandinavian peasants to America that occurred massively throughout the nineteenth century. As Sandberg (2008: 274) further remarks, 'we find that the colonisation was never quite complete; the indigenous groups in the harshest and most marginal regions maintained much of their lifestyle and their customary use of natural resources'.

Transformation of the Sámi Community

At the same time (mid-nineteenth century) the governments were also working toward a novel partition of the Sámi populations, namely into nomadic and sedentary Sámi. It must have seemed easier for the settler states to incorporate those segments into the national populations that had traditionally followed a sedentary way of living or had stopped with nomadic reindeer husbandry by settling at an erstwhile campsite and starting farming. This is the segment that already at this time represented the majority within the Sámi society; moreover, it must be mentioned that nomadism (reindeer pastoralism) among the Sámi started as late as the sixteenth or seventeenth century when communities ceased to live as hunter-gatherers. '[T]he sedentary Sámi were gradually assimilated into the Norwegian, Swedish and Finnish culture from the 1750s onward, by means of national compulsory schooling systems, expanding religious services, agricultural market and support systems, and a cadastre linking property rights to property

10 It was argued by the government that the Sámi were not capable of owning land because of their nomadic way of life.

taxes' (Sandberg 2008: 274–5). In many areas the use of Sámi languages was prohibited in the public sphere and gradually disappeared even from the families' homes. In the long run many people lost their Sámi identity and *truly* became Scandinavians.

This process was stronger in Norway, known as Norwegization, as it was even employed on the nomadic Sámi from the early twentieth century onward. It thus affected the Sámi reindeer herders from the Swedish side of the border: 'When Norwegian nationalism and opposition to the union [with Sweden (1814–1905)] grew stronger at the turn of the century, so did criticism of Sami reindeer herding across the border. Norwegian authorities tried to ward off all kinds of "Swedish penetration," including "the Swedish Lapp traffic"' (Eriksson 1997: 88).

In Sweden, in contrast, nomadic Sámi – at this time the only group of Sámi that was labelled Sámi (or *Lapp*), the others being regarded as already fully assimilated Swedish – became subject to a segregating policy under the motto *Lapp shall be Lapp* (*Lapp skall vara Lapp*) from 1928 onwards (Gunnersdotter 2011: 53). Eriksson (1997: 88) calls this type of segregating policy 'selected segregation'. It has its origin in the formation of a Sámi territory within the Swedish state through the introduction of two borders within Sweden: First, the border of Lapland (*Lappmarksgränsen*) was established in 1766. It demarcated Sámi tax districts. Later, the cultivation border (*Odlingsgränsen*) was created in 1867, further to the west of the earlier border. It served as a tool to prohibit any agricultural activities in an area that was then exclusive to Sámi economic activities (Koch and Miggelbrink 2011: 118, Sandberg 2008: 274, Udtja Lasse 2007: 20). Marainen (1984: 9) compares this territory with an Indian reservation in North America. Interestingly, the policy – *Lapp shall be Lapp* – was based on the romantic image that the (nomadic) Sámi as a wild people are best adapted to the harsh conditions of Arctic Lapland and should not grow soft by Swedish civilization. On the other hand, the Sámi were ultimately bound to a nomadic way of life and as it was equated with a lower state of development, the Sámi were assumed to soon become extinct (Gunnersdotter 2011: 53). Reindeer herding, both the business of herding and the herders as people, were monitored by a special authority, the Lapp police administration (*Lappväsendet*), and until 1971 the herders' children were educated in special schools, so-called tent schools (*kåtaskolor*), that provided no more than basic education, while sedentary Sámi children were sent to ordinary public schools. Eriksson (1997: 88) calls our attention to the very fact that '[i]t is noteworthy that the policy of selective segregation was employed in Sweden alone. In the other partitioning states assimilation was the rule'.

Transformation of Sámi Reindeer Husbandry

The impact of state governance affected reindeer husbandry as radically as the Sámi community in general. 'What were traditionally subsistence activities (for example, reindeer herding, hunting, fishing and so on)', Mazzullo (2010: 103) explains, 'are now carried out more as modern commercial enterprises, usually in combination with extensive involvement in seasonal work, commerce and other tertiary sector activities such as local government and administrative offices, education, social services and tourism'. The rationalization and modernization of reindeer husbandry – especially after World War II when the state societies in general were on the way to welfare capitalism – was regarded necessary since this economic activity is rather inefficient and can hardly ensure an income that households can live from. Rationalization, however, was not only aimed at proficiency. Forrest (2002: 263) for example calls to our attention that '[t]he earliest forms of rationalisation were designed to minimize contact and conflict between herders and settlers. The building of fences, and restrictions on herd sizes, numbers of herders and kinds of herding were used by the states to limit the impact of herding on surrounding agricultural areas'.

When the states started to gain control over husbandry, the first change was to transform the units in which husbandry was organized. Traditionally, reindeer husbandry – and about one thousand years ago all Sámi communities – was organized in social units called *siida*. Each *siida* consisted of several families or households and were provided with its own territory and jurisdiction. From the late nineteenth century when the states started to regulate, control and administer reindeer husbandry, they also thoroughly changed the organization of that activity. In Norway the government created reindeer herding districts (*reinbeitedistrikt*) and subordinated herding units. These now represent individual companies and the government issues licences to start such a company. In Sweden the government installed Sámi villages (*samebyar*) that work similarly to its Norwegian counterparts, but the control over who can work as a herder rests with the local community. Instead of business licences from the state, affiliation therefore is based on membership in the *sameby* which herders usually acquire by birth (Jernsletten and Klokov 2002: 5, Eriksson 2002: 241, Forrest 2002: 254, Rønning 2007: 234). Today, the notion of *siida* still plays a certain socio-cultural role within the communities but is otherwise obsolete in the languages of national law and domestic economy.

Greater changes to the methods of reindeer herding occurred in the 1950s and 1960s when modern technologies were introduced – a process often referred to as the snowmobile revolution (Pelto 1973). This of course made herding more efficient but at the same time it required the husbandry companies to sell more meat and accumulate capital to afford these expensive technologies. 'This duality, the need for cash to purchase new technology and the ability to produce for the cash market made possible by new technology, changed reindeer herding and husbandry from a mainly subsistence economy to a market economy' (Rønning 2007: 235). The increase in meat production, however, had its negative effect as herds grew ever larger in a situation where reindeer husbandry was facing a shrinkage of land (Riseth 2006: 545). It resulted in severe heavy grazing especially in the 1980s, which caused poor pastures conditions (Forbes 2006: 20). This became a massive problem, both in economic and ecological terms. It finally allowed the states – or actually it really required the states – to expand management regulation activities, including resource or land management and paying subsidies (as to any other agricultural business). However, there is also resistance to the advances of state control over Sámi reindeer husbandry and to giving up traditional practices. 'Rationalisation is considered necessary for survival, but the development' Mörkenstam (2002: 124) argues 'is hindered by "a certain conservatism" on the part of the reindeer herders themselves'.

My Sámi, Your Sámi: Regulations of Cross-Border Reindeer Herding

In 1883 the Norwegian and Swedish governments first introduced regulations to cross-border reindeer herding with the Joint Lapp Law (*Felleslappeloven*) which after the dissolution of the Swedish-Norwegian union in 1905 was replaced by border conventions between the then two independent states. The governments were obliged to agree on a new convention every 30 years. Border conventions came into force in 1919, 1949 and 1972. The Norwegian government (assuming the gradual extinction of reindeer husbandry in general) regarded the Swedish-Sámi herders as penetrators of Norwegian territoriality. In negotiations on border conventions with the Swedish government, it enforced the reduction of the so-called convention areas – pastures designated for herders from the other country. The summer pastures in Troms, which were designated to herding communities from the Swedish side, were reduced from 17,000 km² in 1919 to 3,150 km² in 1972 (Map 6.4). Disagreements, among others a confirmation of Swedish-Sámi pastoral rights in the Altevatn area by the Norwegian Supreme Court, were not considered by the negotiating parties, the Norwegian and Swedish governments (Koch and Miggelbrink 2011: 119–20).

During the 1950s and 1960s the Norwegian government gradually changed its opinion about reindeer husbandry and accepted its continued existence. As a response to the great pressures on the pastures in Finnmark caused by the border with Finland being closed until now, the Norwegian government invited Sámi groups to move to what were previously convention areas: areas that the Finnmark Sámi often perceived as *empty areas* since these had not been used for herding for many years. In 1963, the Norwegian government re-arranged the reindeer herding districts in Troms and also allowed Norwegian Sámi to use the remaining convention areas for winter pastures that in the same time period were being used as summer pastures by Swedish Sámi herders. As a result the vegetation of the areas could hardly recover, which meant a decrease in quality of the pastures. Norway clearly transformed from a state trying to discourage people from reindeer herding to a state that started to advocate their own herders. We must, however, recognize that behind the states' obvious affirmation to advocate their respective Sámi reindeer herders' interests in cross-border pasture negotiations stands a societal 'conception of the [Sámi, which] hinges on the notion that they are incapable of handling their own affairs', as Mörkenstam (2002: 114) diagnosed when he dealt with 'the question of self-determination for the Saami people'. He

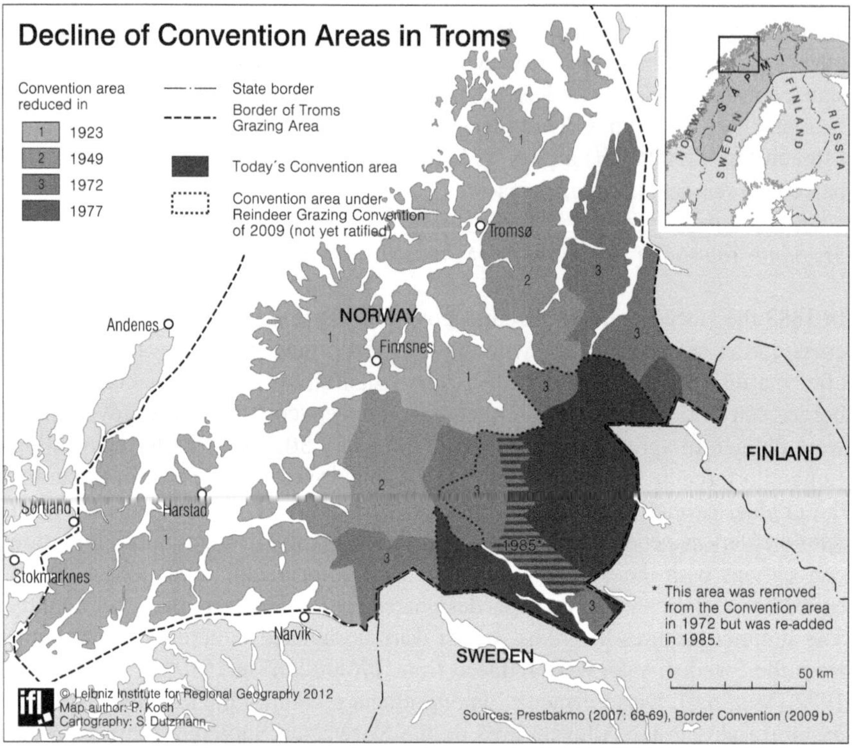

Map 6.4 Decline of convention areas in Troms

further concludes that such a 'conception justifies a paternalistic and "protective" policy'.

To briefly summarize the section on the disciplinary mechanism, I discussed how the states created contents in the territories through policies of colonization, assimilation or segregation, and the integration of Sámi, land and livelihoods. The states became concerned with assigning citizenship and the allocation of pastures. This was possible through the states' establishment of technologies such as administration and education that enabled state control.

Security Mechanism: Towards the States' Acceptance of a Sámi Nation

In this section I will discuss that since the 1980s the Nordic states have started to acknowledge the special position of the Sámi as an indigenous people and Sápmi as their homeland. This includes a political recognition of the Sámi within the framework of each state's political organization and the appearance of Sámi transnationality that goes beyond nation states. At the end of this section I shall outline that this acknowledgement has recently started to affect the management of cross-border reindeer herding between Norway and Sweden.

The Political Recognition of the Sámi as an Indigenous People within the States

During the second half of the twentieth century the states underwent a metamorphosis with regards to dealing with the Sámi populations and territories based on political development and protests by the Sámi and, not least, other segments of society (for example, in the protest against the construction of the Alta hydroelectric power plant). It was the era of the *native revival* in which the United Nations put forward legal standards and international obligations and non-governmental organizations assisted indigenous communities to enforce their rights with political lobbying. In the 1990s the Sámi gained the constitutional status of being an indigenous people in Finland and Norway. In Sweden, state secretary Annika Åhnberg, who represented the Swedish government at the Indigenous Population's Day (*Urbefolkningsdagen*) in Tärnaby on 8 August 1998, officially recognized the Sámi as an indigenous people and apologized for discriminatory policies in the past like the prohibition of using Sámi languages and the forced relocations of Sámi (Udtja Lasse 2007: 45–7). In all three Nordic states with Sámi populations the governments gave way to the installation of Sámi self-determination bodies within the states, known as Sámi Parliaments, which also represents a significant change in the legitimacy basis for political participation of the Sámi as a minority within the state societies (Weigård 2009). These bodies must be consulted by state governments, parliaments, and authorities in any matter that may affect the Sámi communities. Along with this positive development, the situation of Sámi education and the use of Sámi languages in public arenas have improved significantly.

The Appearance of Sámi Trans-Nationality

The changed atmosphere for Sámi self-determination, thus, even allowed a transnational perspective in which the idea of Sápmi (Sámi land) received greater relevance. 'Sapmi is often conceived of as an organic unity between nature and man, land and culture. It represents the unity of an imagined homeland and people. Most importantly, Sapmi is presented as the unity of the Saami as a single people with a single homeland, divided by four separate states' (Eriksson 2002: 240). I think that the idea of Sápmi as a transnational space does not necessarily collide with the nation states' principle of territoriality as the states themselves have developed and institutionalized regionalizations such as an imagined pan-Nordic space, Norden, or the Barents region which also involves Russia. What is specific about Sápmi is that it is *Sámi*.

The Nordic states do not only cooperate cross-nationally on both the governmental and parliamentary level but also support a further strengthening of the Sámi community across its borders with the development of a Nordic

Sámi convention. 'The inspiration for the Saami Convention', Eriksson (2002: 245) argues, 'is the Lapp Codicil from 1751, which recognised the existence of the transnational "Lappish nation"'. Concerning its contents the Nordic Sámi Convention is rather based on the ILO 169[11] – the worldwide most important legal document for indigenous rights – which has thus far been ratified by Norway (Norway was actually the first state that ratified the convention in 1989) but not yet by Finland and Sweden. The ILO 169, however, necessarily must remain abstract with its global inclusion of all indigenous peoples and its general articulation of indigenous rights. What makes the Nordic Sámi Convention relevant, therefore, is the simple fact that it allows a specification of the ILO 169, namely to take the conditions of the Sámi community in the Nordic states into account. Moreover, the Nordic Sámi Convention is a unique legal document because not only the national governments but also the national Sámi parliaments that represent the Sámi populations within each of the states will be signing parties. 'This would imply nothing less than a radically changed definition of international law in the Nordic context, for traditionally only states are primary subjects in international law' (Eriksson 2002: 246).

Besides the notion of a Sámi homeland and a kind-of-territory space for Sámi-Nordic policies, '[i]n the 1990s … Saami activists have increasingly turned to speak of Sapmi as a "region"' (Eriksson 2002: 241). This links Sápmi to EU regional policies, namely the Interreg III A North of which *Sápmi* is a sub-programme aimed at strengthening a region regarded as the periphery to gain regional cohesion within the EU. It promotes cross-border cooperation in the north of Fennoscandia with a focus on Sámi cooperation in cultural and business affairs. Sápmi and the Sámi therefore have become a medium for strengthening supranational action (and not simply the place where EU-induced policies are expedited) (for example, Bauer 2012). Not surprisingly, Karppi (2001: 401) comes to the conclusion that '[i]n the case of the European Union Programme example, the increased space for the Sami was of course a side-product of the EU objectives of downplaying the influence of national borders, and had very little to do with the Sami *per se*'.

Towards a New Cross-Border Reindeer Herding Regime?

As stated at the beginning of the chapter, the most recent convention on cross-border reindeer herding between Norway and Sweden expired in 2005 and has not been renewed, creating conflicts with dire consequences. In the latest proposal of a new convention from 2009, we can observe first signs towards a greater involvement of Sámi reindeer herding groups. In future, local herding groups from both sides of the border shall negotiate themselves when to use which pasture in winter and summer. The states will contribute to it with installing an administrative infrastructure: the Sámi will be supported by two novel joint

11 International Labour Organization, *Convention No. 169 Concerning Indigenous and Tribal Peoples in Independent Countries*.

authorities, a Norwegian-Swedish reindeer herding board (*Svensk-norska renbetesnämnden*) and a Swedish-Norwegian examination board (*Svensk-norska överprövningsnämnden*) (Border Convention 2009a: ch. 5), for example in case of herder disagreements. The new mode of how to deal with cross-border reindeer herding should not be misinterpreted as a greater acceptance of herders' wills by the governments but rather be regarded as a means of withdrawing from direct involvement in pastoral conflicts. This would reduce the state's risk of being held liable and of having a bad reputation in the international sphere. It would also reduce costs to achieve mutually accepted agreements.

In this section I have discussed issues of Sámi–state relations that fall in the logics of the security mechanism. One may have the impression – from my text – that things are turning to the overall positive for the Sámi. This is indeed – at least from my point of view – the case especially in cultural and language matters. We must however necessarily bear in mind that the states have to deal with many different interests in society. This puts the states in a position of managers of these interests and political actions are holistically based on statistical calculations. Through international obligations states must ensure the existence of reindeer husbandry economically because it is culturally important. Another prominent field of politics here is regional development – or sustainable development[12] – where all existing interests found in one region are considered in development plans. It does not only allow but simply demands bottom up activities and lobbying from interest groups like Sámi reindeer herders because they have to compete on resources.[13]

Conclusion

In this chapter I have shown, inspired by Foucault's theory of power mechanisms, how the states have treated the Sámi, their homeland and reindeer herding throughout the last three centuries. The overall result of this study is the identification of two reasons why cross-border reindeer herding between Norway and Sweden still exists today. First, with the juridical mechanism at work, the states established exclusive territories through bordering, which clarified the extent of the states' sovereignty: land, people, goods etc. By and large, this ended the multiple taxation of the Sámi. Since the implementation of state territoriality deeply collided with Sámi spatiality, it provoked a new problem – cross-border reindeer herding – something the states could not deal with at that time. They simply omitted this issue, and to demonstrate their capacity to act (and not to lose face), they explicitly permitted the free migration of Sámi herders across

12 *Sustainable development* appears to me, personally, as a summary of the security mechanism in two words!

13 In most cases, this may be a rather difficult task to indigenous communities and locals in general. See Mazzullo's chapter in this volume, especially the section on maps and space-making (to be regarded as technologies).

the border in the Lapp Codicil. This legal document, therefore, must be clearly regarded as an outcome of the states' weakness to enforce full control over the newly demarcated state territories in the mid-eighteenth century rather than the states' strength and good will to recognize the traditional rights of the Sámi, as many accounts would have it. However, this does not question the Lapp Codicil's central importance – stylized as the Sámi Magna Carta – for Sámi–state relations and the transnational Sámi community.

Second, with the disciplinary mechanism at work, the states started to transform the Sámi, the new territories and reindeer husbandry within the juridical frameworks of each state. In most cases the Sámi and their territories were assimilated into the national societies, that is the states corrected the Sámi to become state citizens. Many Sámi lost any notion of Sámi identity over time. Only reindeer-herding mountain Sámi in Sweden were segregated from the overall Swedish society in a special territory of Lapland and were subjected to a set of policies aimed at ensuring the Sámi to be perfectly adapted to the romantically perceived *wilderness* of Lapland, that is, the state perfected these Sámi to be *real* Sámi. These policies not only excluded the Sámi from Swedish society but also Swedish settlers to start farming in Lapland. Selective segregation in Swedish Lapland may also explain why Swedish Sámi today seem to have stronger pastoral rights across the border. Reindeer herding was transformed from a holistic livelihood to a profitable meat producing industry within the overall national economies subjected to state administration and supervision. Modern technologies and forms of organization were introduced with the aim of increasing rationalization and the efficiency of this business, and pastures became the most important economic resource of reindeer herding. The states obviously succeeded with correcting reindeer herding and consequently started to advocate *their* respective herders to ensure their economic interests (that is, pastoral rights to specific herding areas). The effect was that herding areas for Swedish Sámi in Norway significantly decreased, although cross-border reindeer herding never ceased.

With the security mechanism at work, the states – and recently the European Union too – have started to treat the Sámi and Sápmi as a nation both internally and trans-nationally for the last three decades. The Sámi nation and Sápmi as its quasi-territory can be regarded as a milieu with a set of natural and artificial givens of *Sáminess* in northern Fennoscandia. The Sámi have been recognized as (the European Union's only) indigenous people and Sámi parliaments as self-determination institutions have been established according to the model of Western parliaments. A set of policies have thus made a transnational perspective for the Sámi possible that goes beyond nation state borders. There are first signs toward a new treatment of cross-border reindeer herding between Norway and Sweden. The two states seem to create a situation in which reindeer herders are supposed to negotiate pastoral rights themselves in the future. The states obviously seek to prevent the risk of a bad image and to minimize the costs caused by negotiations. Whether the Sámi herders will accept the states' new policy of partial disappearance, which includes great challenges to adapt to new socio-political conditions, is hard to forecast. Ironically it only implies a strengthening of the Sámi or an indigenous identity (as opposed to the citizenship-based identity) among these reindeer herders who are practically Sámi by Norwegian and Swedish husbandry law. However this will require legal conditions that allow herders to identify themselves as Sámi (rather than Norwegian Sámi or Swedish Sámi) even when it comes to issues of cross-border reindeer herding. These legal conditions can only be achieved through a harmonization of national reindeer husbandry laws by the states.

Acknowledgements

I thank Judith Miggelbrink and Joachim Otto Habeck for reviewing and discussing my chapter.

References

Bauer, L. 2012. *Sámi Land in Context of the European Union's Multiscalar Metagovernance: A reconstruction and mapping of the Interreg IIIA-Nord Subprogramme Sápmi*. Diploma Thesis, University of Leipzig, Leipzig.

Border Convention 2009a. *Konvensjon mellom Norge og Sverige om grenseoverskridende reindrift – Konvention mellan Sverige och Norge om gränsöverskridande renskötsel* [Convention between Norway and Sweden on cross-border reindeer herding]. [Online]. Available at: http://www.regjeringen.no/upload/LMD/Vedlegg/Regelverk/Reindrift_konvensjon_Norge_Sverige_260209.pdf [accessed: 30 October 2012].

Border Convention 2009b. *Områdeprotokoll til konvensjon mellom Norge og Sverige om grenseoverskridende reindrift – kartvedlegg 1/Områdesprotokoll till konvention mellan Sverige och Norge om gränsöverskridande renskötsel kartbilaga 1* [Area minute of the convention between Norway and Sweden on cross-border reindeer herding – map appendix 1]. [Online]. Available at: http://www.regjeringen.no/upload/LMD/Vedlegg/bilder_illustrasjoner/Reindrift_omrade1_kor_m.pdf [accessed: 30 October 2012].

Eriksson, J. 1997. *Partition and Redemption – A Machiavellian Analysis of Sámi and Basque Patriotism*. Umeå: Department of Political Science, Umeå University.

Eriksson, J. 2002. The construction of Sápmi: Towards a transnational polity? in *Conflict and Co-Operation in the North*, edited by K. Karppi and J. Eriksson. Umeå: Norrlands Universitetsförlag i Umeå, 239–50.

Finnish Government, 2007. Regeringens proposition till Riksdagen med förslag om godkännande av vissa ändringar av gränslinjen som konstaterats vid 2006 års översyn av riksgränsen mellan Finland och Sverige [Proposition of the government to the Finnish parliament with a proposal to acknowledge some changes of the border line that have been stated at the 2006 examination of the Finnish-Swedish border]. RP 128/2007 rd. [Online]. Available at: http://www.finlex.fi/sv/esitykset/he/2007/20070128.pdf [accessed: 17 January 2012].

Forrest, S. 2002. The territorial dimension of Saami-state politics, in *Conflict and Co-Operation in the North*, edited by K. Karppi and J. Eriksson. Umeå: Norrlands Universitetsförlag i Umeå, 251–67.

Foucault, M. 2007 [1977/78]. *Security, Territory, Population: Lectures at the Collège de France, 1977–78*, edited by M. Senellart. London: Palgrave Macmillan. [Online]. Available at: http://www.palgrave.com/PDFs/1403986525.pdf [accessed: 31 October 2012].

Forbes, B. 2006. The challenges of modernity for reindeer management in northernmost Europe, in *Reindeer Management in Northernmost Europe: Linking Practical and Scientific Knowledge in Social-Ecological Systems*, edited by B. Forbes, M. Bölter, L. Müller-Wille, J. Hukkinen, F. Müller, N. Gunslay and Y. Konstantinov. Berlin: Springer, 11–25.

Gruda, L. 2007. *Norway and Sweden Struggle about Sámi Rights: Merely a Conflict on Reindeer Pasture or Deliberate Discrimination of the Indigenous People?* [Online]. Available at: http://www.gfbv.it/3dossier/eu-min/sami-en.html [accessed 31 March 2011].

Gunnersdotter, Y. 2011. 'Det blir bara värre om man inte gör upp med det gamla': Utmaningar för jakt- och fisketurismen i de svenska fjällen ['It just becomes worse if one does not overcome the old': Challenges for the hunting and fishing tourism in the Swedish fells], in *Uppsala mitt i Sápmi – Rapport från ett symposium arrangerat av Föreningen för samiskrelaterad forskning i Uppsala, Upplandsmuseet 4–5 maj 2011*, edited by H. Tunón, M. Frändén, C.-G. Ojala and M.-B. Öhman. CBM:s skriftserie, 55: 52–6 [Online]. Available at: http://www.slu.se/Global/externwebben/centrumbildningar-projekt/centrum-for-biologisk-mangfald/Dokument/publikationer-cbm/cbm-skriftserie/skrift55.pdf [accessed: 29 October 2012].

Hansen, L.I. 2005. Fra Nöteborgsfreden til Lappekodisillen, ca. 1300–1751: Folkegrupper og statsdannelse på Nordkalotten med utgangspunkt i Finnmark [From the Peace of Nöteborg to the Lapp Codicil, ca. 1300–1751: Peoples and state building in the North Calotte with a starting point in the Finnmark], in *Grenser og grannelag i Nordens historie* [Borders and neighbours in Nordic history], edited by S. Imsen. Oslo: Cappelen Akademisk Forlag, 362–86.

Hansen, L.I. 2010. Tax, trade and diversification among the Sami during the late 16th century, in *Transference*, edited by W. Østreng. Oslo: CAS, 1–11. [Online]. Available at: http://www.cas.uio.no/Publications/Seminar/0809Hansen.pdf [accessed: 13 December 2011].

Hedström, J. 2011. Hur länet Norrbotten kom till? I fjol, våren 2010, var det 200 år sedan Norrbotten blev eget län [How did the county of Norrbotten come about? Last year, spring 2012, it was 200 years after Norrbotten became a county]. *Lulebygdens forskarförening medlemstidning*, 74, 8–12. [Online].

Available at: http://lulebygden.se/tidning/Tidning-LF-74.pdf [accessed: 17 January 2012].

Heikkilä, L. 2006. *Reindeer Talk – Sámi Reindeer Herding and Nature Management*. Acta Universitatis Lapponiensis, 110. Ph.D. dissertation, University of Lapland, Rovaniemi.

Ingold, T. 1980. *Hunters, Pastoralists and Ranchers: Reindeer Economies and Their Transformations*. Cambridge: Cambridge University Press.

Jernsletten, J.-L. and Klokov, K. 2002. *Sustainable Reindeer Husbandry*. Tromsø: Centre for Saami Studies, University of Tromsø.

Karppi, K. 2001. Encountering different territorialities: Political fragmentation of the Sami homeland. *Tijdschrift voor Economische en Sociale Geografie*, 92(4), 394–404.

Koch, P. and Miggelbrink, J. 2011. Being in the frontline of a Sámi culture and a private business: Cross-border reindeer herding in northern Norway and Sweden. *Nomadic Peoples*, 15(1), 114–43.

Lähteenmäki, M. 2006. *The peoples of Lapland: Boundary demarcations and interaction in the North Calotte from 1808 to 1889*. Vaajakoski: Gummerus Kirjapaino.

Marainen, J. 1984. *Förflyttningar i Sápmi: Gränseregleringar i Norden och deras Följdverkningar för Samer* [Relocations in Sápmi: Border regulations in the North and its consequences for the Sámi]. Umeå: Svenska Samernas Riksförbund.

Mazzullo, N. 2009. Sápmi: A symbolic re-appropriation of Lapland as Saamiland, in *Máttut – Máddagat: The Roots of Saami Ethnicities, Societies and Spaces/ Places*, edited by T. Äikäs. Oulu: Giellagas Institute, University of Oulu, 174–85.

Mazzullo, N. 2010. More than meat on the hoof? Social significance of reindeer among Finnish Saami in a rationalized pastoralist economy, in *Good to Eat, good to Live with: Nomads and Animals in Northern Eurasia and Africa*, edited by F. Stammler and H. Takakura. Sendai: Center for Northeast Asian Studies, Tohoku University.

Mörkenstam, U. 2002. The power to define: The Saami in Swedish legislation, in *Conflict and Co-Operation in the North*, edited by K. Karppi and J. Eriksson, Umeå : Norrlands Universitetsförlag i Umeå, 113–45.

Omma, I.A. 2006. *Samisk sedvanerätt enligt Lappcodicillen och renbeteskonventionerna: Har rätten förändrats?* [Sámi customary rights according to the Lapp Codicil and reindeer herding conventions]. Umeå: Umeå Universitet.

Pedersen, S. 2002. Norwegian nationalism and Saami areas as no-man's land, in *Conflict and Co-Operation in the North*, edited by K. Karppi and J. Eriksson. Umeå: Norrlands Universitetsförlag i Umeå, 167–82.

Pelto, P.J. 1973. *The Snowmobile Revolution: Technology and Social Change in the Arctic*. Menlo Park, CA: Cummings.

Prestbakmo, H. 2007. *Bardu og Målselv – "østlappenes" land?* [Bardu and Målselv – the land of the eastern Lapps?] Evenskjer: Skániid girjie.

Riseth, J.Å. 2006: Sámi reindeer herd managers: Why do they stay in a low-profit business? *British Food Journal*, 108 (7), 541–59.

Rønning, L. 2007. Entrepreneurship among Sámi reindeer herders, in *International Handbook of Research on Indigenous Entrepreneurship*, edited by L.-P. Dana and R.B. Anderson. Northampton: Edward Elgar Publishing Ltd, 232–45.

Sandberg, A. 2008. Collective rights in a modernizing North – on institutionalizing Sámi and local rights to land and water in northern Norway. *International Journal of the Commons*, 2(2), 269–87 [Online]. Available at: http://www.thecommonsjournal.org/index.php/ijc/article/download/41/25 [accessed 6 January 2012].

Simowitsch, J. 2010. *Flächennutzungskonflikte der Sámi in Norrbotton und Troms* [Land use conflicts of the Sámi in Norrbotten and Troms]. Diploma Thesis, University of Leipzig, Leipzig.

Storm, D. 2009. Könkämävuomasamenes flytinger til sommerområdene i Troms fylke – en forskningshistorisk oversikt [Migrations of the Könkämävuoma-Sámi to summer areas in the county of Troms – a history of science overview], in *Könkämävuoma-samernas renflyttningar till Norge: Om sommerbosättningar i Troms fylke på 1900-talet*, edited by L.J. Walkeapää. Tromsø: Tromsø Museum – Universitetsmuseet, 13–44.

Tuchtenhagen, R. 2009: *Kleine Geschichte Norwegens* [Short history of Norway]. Munich: Beck.

Udtja Lasse 2007. *Begrav mitt hjärta vid Udtjajaure* [Bury my heart at Udtjajaure]. Stockholm: Emma Publishing.

Urbańczyk, P. 1992. *Medieval Arctic Norway*. Warsaw: Semper.

Weigård, J. 2009. Legitimitetsgrunnlaget for samiske rettigheter [The legitimacy basis for Sámi rights]. *Tidsskrift for samfunnsforskning*, 50(1), 27–54.

Chapter 7
Identity Categories and the Relationship between Cognition and the Production of Subjectivities

Brian Donahoe

Serious engagement with cognition and cognitive approaches in anthropology can be traced back to the mid-1950s,[1] with the development of 'ethnoscience' studies at Yale University and further analyses of the semantics and taxonomies involved in native systems of categorization. The field steadily developed over the decades in both breadth and depth to include decision models, consensus mapping, computer-aided discourse analysis, and studies of the cognitive aspects of emotions and religion (D'Andrade 1995, Colby 1996). In the course of its development, cognitive anthropology earned a reputation as one of the most methodologically rigorous of the subdisciplines of sociocultural anthropology. However, despite (or perhaps because of) their highly developed methods,[2] cognitive approaches have always been somewhat marginal within the discipline as a whole. This is still the case today. In fact, some cognitive approaches are rather vilified by many sociocultural anthropologists. This is especially true of those approaches that fall under the rubric of evolutionary psychology, which is often caricatured as 'biological determinism' and accused of trying to reduce cultural phenomena such as religious concepts and behaviour to a specific gene or some module that can be identified and located in the brain. This resistance is, of course, reinforced by abuses of biological and evolutionary explanations (for example, genetic determinism and/or the assertion of the superiority of certain 'races') and an almost tyrannical political correctness within the social sciences that closes the door on certain types of research.

Finally, sociocultural anthropologists have understandably preferred to adhere to what can be observed – public symbols and forms, people's behaviours, the *results* of decisions (for example, revealed preferences) – and to avoid speculating

1 I do not consider the 'culture and personality' school (best exemplified by Benedict 1934) to be a genuinely cognitive approach. In my opinion, that school falls under the rubric of psychological anthropology.

2 Colby (1996: 209) goes so far as to suggest that '[s]ince these analytical approaches are rarely, if ever, found in such subfields of social anthropology as psychological anthropology or interpretive anthropology, cognitive anthropology is more easily conceptualized as distinct from social anthropology rather than as a subcategory of it'.

about what happens in the black box of the brain, i.e., the cognitive and mental states and processes involved in interpreting symbols and forms, motivating behaviour, and making decisions. This preference for the observable was the driving principle behind behaviourism, the dominant metatheory in psychology until the 1950s, but it was supplanted (in the United States and Great Britain) by cognitive approaches following the 'cognitive revolution' in the 1950s and 1960s (Baars 1986, Gardner 1987).

All of these reasons for avoiding overtly cognitive approaches can be supported by 'bad' (i.e., naïve, reductive, over-simplified, overly deterministic) examples of cognitive approaches, or by caricatured readings of more nuanced uses of it. But cognitive and brain sciences are making great strides forward and, as noted by DiMaggio (1997: 266), 'cognitivists have developed ingenious empirical techniques ... that permit strong inferences about mental structures, going far toward closing the observability gap between external and subjective aspects of culture'. Such new techniques and advances, a couple of which I shall refer to later in this chapter, make it possible to engage with these sciences in more sophisticated ways and to combine them with sociocultural anthropological approaches. This will not happen, however, until anthropologists make a genuine effort to engage with the cognitive sciences. Cognitive anthropology is still on the fringes of the discipline,[3] and while there is no need for everyone to practice it, there is also no need to vilify it. Moreover, given the central position and dominance within sociocultural anthropology of certain concepts such as identity, subjectivity, agency, imaginary (as a noun), the self, and consciousness – all of which are in fact terms laden with psychological and cognitive connotations and assumptions – anthropologists would be well advised to acknowledge the psychological and the cognitive dimensions implicit in such terms just to make sure they are using them correctly. As Claudia Strauss and Naomi Quinn put it in the introduction to their seminal volume *A Cognitive Theory of Cultural Meaning* (1997):

> Nowadays in cultural anthropology it is perfectly respectable to talk about 'the self,' 'meaning,' 'identity,' 'consciousness,' 'subjectivity,' 'experience,' 'reader response,' 'the imagined,' and 'agency'; strangely, however, we are not supposed to talk about the psychological processes and structures that help explain these. If we ignore psychology, however, we are likely to make false assumptions about the ways selves, meanings, identities, consciousness, subjectivity and experience are constructed and about the way people respond to texts, imagine communities, and resist hegemonic structures. (Strauss and Quinn 1997: 9, see also Quinn and Strauss 2006)

Just as psychologists were forced by new insights and advances to shift the paradigm from behaviourism to cognition in the 1950s, so anthropologists who

3 This is less so the case in French anthropology.

rely in any way on psychological concepts, as most do, should also work towards making that shift.

An important point of clarification is in order here. The term *cognitive* is ambiguous. It is often used to denote 'conscious, intellectual functions' (Baars 1986: 5), as is implied in the verb *cognize*, but I (following cognitive sciences) take it to mean something much broader that includes all the workings of the human brain and its interactions with the external environment, both social and physical.[4] The great majority of these functions remain below the surface of consciousness.

Categories, Classification and Subjectivities

Identity, broadly speaking, has arguably been the central focus of the discipline of sociocultural anthropology for much of the past thirty years. Anthropologists often invoke 'the cognitive' in their analyses of identity formation and identification, relying on many of the concepts mentioned by Strauss and Quinn in the above-quoted excerpt. However, as I will show below, these anthropological treatments of identity seldom go beyond a rather superficial invocation of cognition and 'the cognitive'. In this chapter I aim to take some small steps towards entering the black box of mind, brain, and cognition by relating cognition to the formation of identities and subjectivities, and specifically to indigenous identities and subjectivities, and to the broader general themes of this volume: the production and cognition of spaces and subjectivities, and the cognitive dimensions of spaces and subjectivities.

While the concept of subjectivity can be traced back to Hegel and Kant, it has entered contemporary anthropology primarily through the work of Foucault and his discussions of how governmentality – various technologies of governance – produces subjectivities, meaning 'the way in which the subject experiences himself in a game of truth where he relates to himself'.[5] It is an agent-centred position of identity in relation to particular forms of knowledge, power, and practice. One of the ways subjectivities are produced is through state-created categories and classifications that people come to identify with, and in relation to which people start to think and act in new ways. Such categories and classifications include environmental (Agrawal 2005), indigenous (Donahoe

4 There has been a great deal of debate over the use of the terms 'brain' and 'mind', the relationship between them (see, for example, Damasio 2010), and the degree to which the two may or may not be interchangeable. The social sciences tend to prefer the term 'mind'. In this chapter I use 'brain', as I am focusing on the neurobiological processes that, I suggest, underlie various subjective experiences of identity and the formation of subjectivities.

5 From 'Foucault', an entry in the *Dictionnaire des philosophes* (1984, 942–4) written by Foucault about himself, but submitted under the pseudonym Maurice Florence.

2011), aboriginal, native, nomadic, and tribal, as well as any number of more specific ethnic-group labels.

The widely accepted assumption within the social sciences is that such categories and classifications are constructed, and that they are generally not constructed by the people to whom they refer. As Fredrik Barth (1969) pointed out so many years ago, people do not tend to define themselves and their culture; rather they tend to define the boundaries that set them apart from other people. Until there is a need to do so, people do not think of themselves in such terms as indigenous, native, nomadic, etc. These categories are constructed by others who often fulfil functions within state structures. In this way, such categories are an integral tool in the technologies of state-building and governance, including the categorization and enumeration of populations for purposes of effective administration (for example, censuses).[6]

In the case of indigenous peoples today, who are by all (externally constructed) definitions of the term, marginalized, non-dominant, politically and economically subordinated to a dominant majority, the terminology and categories (indigenous, native, aboriginal, tribal, etc.) are created for them and it is then up to them to accept and work within this terminological framework (as Russia's indigenous peoples for the most part do[7]), modify it (as Canada's First Nations have done), or reject it outright. In some cases, there has been an initial resignation, then a rejection of the terms of reference, and finally an almost militant appropriation of the terms (for example, the use of the term *Indian* in the United States).[8]

Over time, these categories and classifications (often with modifications) are transmitted through processes of enculturation and socialization and become a type of collective representation, to use Durkheim's phrase (1965 [1912]). Moreover, the processes of engaging with such identity categories – whether one accepts those identities uncritically, resists them vehemently, plays along with them to take advantage of them, or whatever in-between combinations of such responses are possible – are definitely in the realm of cognition.

Classification and categorization, and the 'interplay between self-identification and external categorization' (Brubaker, Loveman and Stomatov 2004: 32) figure prominently in numerous studies of ethnicity and identity more generally, and in

6 See Appadurai (1988, 1998), Barrier (1981), Cheshko (2005 [2000]), Cohn (1987), Hacking (1982), Hirschman (1987), Kertzer and Arel (2002), Porter (1995), Sokolovski (2005a), Sokolovskii (2005b [2000]), Stepanov (2005 [2001]) and Urla (1993).

7 For an example of how an indigenous group in Russia strives to conform to the state's terms of reference, see Donahoe (2011).

8 See, for example, the comment, 'I am an American Indian, not a Native American', by Russell Means, Oglala Sioux activist and first national director of the American Indian Movement (available at: http://web.archive.org/web/20090503130744/http://www.peaknet.net/~aardvark/means.html; accessed 8 November 2012).

many of these cases cognition is either implied or superficially invoked.[9] In fact, I was involved in one such study, a very protracted exercise in corporate authorship of a theoretical paper on identity formation (Donahoe et al. 2009). One of the big sticking points was whether we should attempt to get into the black box of the mind or not. The project was thorny enough, and in the end we decided not to tackle the actual cognitive processes such as bounded rationality, how identity formation taps into evolutionarily formed predispositions, biases, and heuristics, and the role of emotions from a cognitive perspective, all of which would have made the project even thornier and virtually intractable.

I recently came across an interesting discussion paper by Dominik Helling (2009) that promised to go a bit more deeply into the cognitive aspects of identity formation, or nation-building, which is the more specific focus of Helling's paper. Helling's main thesis is that nation-building is an exercise in what he calls 'socio-cognitive standardisation', which he defines as 'a process of identity formation underwritten by the spread of social and cognitive conformity within a politically and usually geographically defined population. Socio-cognitive standardisation is understood as a procedure whereby one set of socio-cognitive elements – such as language, history or mental maps – becomes a society's chief common framework of reference' (Helling 2009: 1). Helling goes on to say that 'central state administration is the prime driver for the process of socio-cognitive standardisation that underlies nation building', (Helling 2009: 1–2) because 'cultural identities (i.e., ethnicities) are defined by institutional design' (Helling 2009: 11). Somewhat disappointingly, however, Helling concludes that 'it is of secondary importance how this socio-cognitive standardisation is brought about' (Helling 2009: 22).

In a 1994 paper called 'Rethinking the state: genesis and structure of the bureaucratic field' by Pierre Bourdieu, Loïc Wacquant and Samar Farage the authors set out to address exactly this question of how such socio-cognitive standardization is brought about. The authors note that '[i]f the state is able to exert symbolic violence, it is because it incarnates itself simultaneously in objectivity, in the form of specific organizational structures and mechanisms, and in subjectivity in the form of mental structures and categories of perception and thought' (Bourdieu, Wacquant and Farage 1994: 3–4). They go on to assert that

> [t]hrough classification systems (especially according to sex and age) inscribed in law, through bureaucratic procedures, educational structures and social rituals ... the state molds *mental structures* and imposes common principles of vision and division.... And it thereby contributes to the construction of what is commonly designated as national identity. (Bourdieu, Wacquant and Farage 1994: 7–8, emphasis in the original)

---

9 See, for example, Banks (1996), Cornell and Hartmann (1998), Erikson (1993) and Jenkins (1997).

The authors conclude that 'acts of submission and obedience' to such symbolic violence are 'cognitive acts which as such involve cognitive structures, forms and categories of perception, principles of vision and division' (Bourdieu, Wacquant and Farage 1994: 12–13). Once again, the actual processes of cognition are assumed but not addressed in this treatment linking state formation to cognition. Bourdieu's concept of *habitus*, which clearly has cognitive origins and is highly influenced by the work of Jean Piaget,[10] underpins the argument in this study, but Bourdieu himself never gets very specific or precise about the actual cognitive underpinnings of *habitus*.

One recent overview, 'Ethnicity and cognition' by Rogers Brubaker, Mara Loveman and Peter Stamatov (2004), promises a more concrete connection between categories, ethnicity and cognition. The authors see 'the emergent concern with categorization as an incipient, and still implicit, cognitive turn in the study of ethnicity' (Brubaker, Loveman and Stamatov 2004: 31), but note that 'the scope of this cognitive turn … has been limited by the lack of engagement with expressly cognitive research in psychology and cognitive anthropology' (Brubaker, Loveman and Stamatov 2004: 36). Brubaker and his co-authors note that official categories that are imposed through censuses, laws, and other administrative actions 'have profound consequences for the self-understandings of the classified', but that these categories 'differ substantially' from the 'categories used by ordinary people in everyday interaction' (Brubaker, Loveman and Stamatov 2004: 35). This suggests two different ways of cognizing identity that can exist simultaneously, and it would be an interesting if challenging project to try to understand how these operate cognitively.

None of these approaches engages seriously with cognition as processes that address the mutually constituting dynamics between brain, body, and the environment, both natural and social, and they do not engage with recent advances in what may be termed brain sciences or cognitive sciences. Arguably, this is generally the case in most sociological and sociocultural anthropological studies that invoke cognition in one form or another.

At a cognitive level, these categorizations and classifications tap into two evolved human propensities and needs: the need to classify and categorize in order to make sense of the world (see Ellen 2006); and the social need to belong to a group. As George Lakoff noted, 'There is nothing more basic than categorization to our thought, perception, action, and speech' (cited in Brubaker, Loveman and Stamatov 2004: 37). But how can we understand what is going on at the level of cognition in the heads and bodies and communities of the indigenous, tribal, nomadic peoples with whom we work? What happens *cognitively* to people when, for example, they dress their own thoughts in new language, or they present their own experiences in terms different from those they would normally express them in, just to meet certain expectations, to fit certain externally constructed and imposed categories? What effect does this new way of representing themselves linguistically and

10 See Lizardo (2004), see also Strauss and Quinn (1997: 44–7).

conceptually have on their cognitive processes? How are such externally imposed constructed categories internalized and even embodied, how are they related to the production of space and place, and how do they get drawn into the delicate cognitive dance of brain-body-environment? Much of the literature on indigenous and nomadic experiences of the land and environment emphasize human agency in shaping the landscape, and active engagement in the environment as a crucial factor in developing the more embodied phenomenological 'emplacement in the world', to use one of Ingold's (2000: 144) phrases. This is no doubt the case, and the knowledge and the skills that a person must master to live as one with one's natural surroundings certainly shape cognition and even the physical structure of the brain in certain ways.

Similarly, I would suggest that the politics of identity and recognition is a field of practice, and active and agentic participation in it can also change people's cognition. This field of practice is also a type of social environment, and the more convincingly one can engage with it, the more genuinely one can embody a certain category and emplace oneself in the world of categories and classifications, the better one's chances of success will be within this environment. As Bourdieu, Wacquant and Farage put it,

> cognitive structures are not forms of consciousness but *dispositions of the body*. … Submission to the established order is the product of the agreement between, on the one hand, the cognitive structures inscribed in bodies by both collective history (phylogenesis) and individual history (ontogenesis) and, on the other, the objective structures of the world to which these cognitive structures are applied. State injunctions owe their obviousness, and thus their potency, to the fact that the state has imposed the very cognitive structures through which it is perceived. (Bourdieu, Wacquant and Farage 1994: 14, emphasis in the original)

The link between identity and identity categories, on the one hand, and the material and symbolic production and cognition of place and space on the other hand, becomes more apparent when we remember how both phenomenological experience and state categorizations of place and space are implicated in identity formation and production. In the Russian Empire and the Soviet Union, for instance, the definition of *indigenous* has always been linked to residence within a certain territory. Russian and Soviet administrators have long applied a 'strategy of linking peoples to territories and via territories to rights' (Sokolovskii 2005a and 2005b: 17). This relates to the presumed existence of ethnic homelands and what Russian anthropologist Sergei Sokolovskii (2005a) has termed 'territorialized ethnicity'. More recently in Russia this territory-ethnicity linkage has been carried over and incorporated into the official definition of 'indigenous, small-numbered peoples of the Russian Federation' (*korennye malochislennye narody Rossiskoi Federatsii*), which stipulates that living in 'territories of traditional settlement of their ancestors' is one of the necessary conditions of eligibility for some of the privileges and concessions that go along with indigenous status. Of course at this level 'territory' is not the same as place, and certainly not the same as space. At best it is a weak proxy measure for the experiential dimensions of the attachment to a place that comes with moving through a place or places as space. Nevertheless, over time people form strong attachments to these state-assigned territories, just as they do to imagined communities. The point is that for indigenous peoples to make these state-created legal concepts and categories work at all for them, they have to frame their claims in those terms, and over time this inevitably starts to have an impact on their actual perceptions and cognitions of place and space.

Intimations of Methodology

Now the really important and tricky question is the issue of methodology. Language is assumed to be an important window into cognition, and as such it has been the tool of choice for researchers to get at the way people cognize their surroundings (for example, Conklin 1954, 1980, Berlin, Breedlove and Raven 1966, Berlin and Kay 1969). That is the fundamental assumption underlying ethnoscience and its associated methods (componential analysis, etc.). It is also what all the wailing and gnashing of teeth over the loss of linguistic diversity is all about (see, for example, Harrison 2007). Such linguistic-based approaches have an underlying Whorfian assumption about the relativity of language – that language shapes thought and cognition to a greater or lesser degree. But how much water does this assumption hold?

Specifically with regard to spatial conceptualization, which is directly related to the cognition of spaces and places, Stephen Levinson and his colleagues (most notably John Haviland) have done some interesting work comparing absolute and relative systems of spatial conception, using Guugu Yimithirr speakers in North Queensland and Dutch speakers. In absolute systems, directions are fixed and not relative to the speaker; they do not vary with body rotation (for example, cardinal directions). In relative systems, on the other hand, angles and directions are relative to the speaker's point of view, such as back, front, left, right, etc. Linguistically, Guugu Yimithirr speakers' system of spatial description lacks relative notions such as 'front', 'back', 'left' and 'right', instead using absolute terms similar to cardinal directions, whereas Dutch speakers (and other Indo-European languages) rely more heavily on relative notions. Levinson and his colleagues were interested in seeing if this linguistic difference reflected an actual cognitive difference as well, and devised a series of tests to see if Guugu-Yimithirr speakers actually *think* in these more absolute terms as well. Without going into the details, it turns out that they do, and Levinson concludes that this is 'as good an example of "linguistic relativity" as one might hope for: a linguistic (and more broadly communicative) system may require one to think in a certain distinctive way' (Levinson 1998: 20).

Another particularly productive semantic domain with regard to analysing cognition is toponyms. As with the loss of linguistic diversity more generally, we can ask: What is actually lost when place names are forgotten? As Keith Basso, the author of one of the most famous examples of the value of place names, has put it: 'Less often still has placenaming been investigated as a universal means – and it could well turn out, a universally primary means – for appropriating physical environments' (Basso 1996: 76). Work on Native American place names demonstrates, first, how differently place names are constructed in those languages when compared to English and other Indo-European languages, and second, how the syntactic structure of the languages would appear to reflect a different way of cognizing the landscape. Thomas Thornton in particular, in his 2008 book on Tlingit place names, compares Tlingit and English names for the same places in Glacier Bay National Park in Alaska. English place names are more often than not made up of a 'classic binomial compound, consisting of a generic physical feature of the landscape (bay) with a specifier, in this case a noun (glacier), preceding it' (Thornton 2008: 81). Thornton also noted that fully 50 per cent of the English place names he collected in Glacier Bay National Park were biographical, named after statesmen or explorers, etc., while only one per cent of the Tlingit names were biographical, and in fact only one per cent of all native toponyms in southeast Alaska were biographical. While the Tlingit place names also often included a generic feature of the landscape, these were paired with verbal constructions that described the places in terms of actions, motion, and processes. Tlingit, like Basso's Western Apache and many other native American (and aboriginal Australian and Turkic and other) languages, allow complex descriptive phrases to be built out of a single verb stem using prefixes, infixes and suffixes, so a great deal of process and action can be very economically built into rather short phrases. In place names the possessive suffix figures prominently, not to indicate that a certain person owns that place, but rather to indicate that the place possesses things, including people. Thornton convincingly demonstrates that these place names are indicative of Tlingits' cognition of place and space, and concludes that 'three important cognitive orientations that are embedded in place can be identified: social identity, time-space, and normative orientations' (Thornton 2008: 104).

What Roland Barthes (1996 [1982]) has called a 'semiology of landscape' must go beyond the question of linguistic structure to include a history of significant social events and actions in the environment. Such significant social events can get built into place names (as Basso and Thornton have shown); they become depositories of stories, histories, and morals; they are part of social and cultural memory, and as such, enter into the cognition of individuals. But how do they enter into cognition? In his 2002 book *Connections*, anthropologist Stephen Reyna develops what he calls a 'Boasian cultural neurohermeneutic system'. Reyna's book is the most thorough and sustained exercise in engagement with cognitive sciences by a sociocultural anthropologist that I have yet come across. According to Reyna, the neurohermeneutic system interprets and connects antecedent events that have happened either in one's personal history or in a shared history to form the basis of subsequent actions and events. It does this via a 'connector', which is composed of linked neural circuits in the brain. I suggest that place names can trigger just such a connection by bringing antecedent events to bear on a present situation to inform subsequent action.

In any case, linguistic methods have probably been the most common way of trying to get at actual cognitive processes. There are, however, other possibilities. In an earlier paper (Donahoe 2009) I suggested that significant events such as social dramas or environmental disasters, i.e., 'events generated by the human (social, economic, or political) environment that have environmental components and/or repercussions'[11] (Lees and Bates 1990: 266), could be analysed using concepts and findings from game theory – something akin to 'games in the wild' (compare Hutchins's *Cognition in the Wild*, 1995). Such an event-focused approach 'offers a convenient "point of entry" for the description of complex and changing relationships' because it forces the data into a diachronic perspective (Lees and Bates 1990: 255). In doing so, events can help bring routine, everyday 'automatic cognition' to the surface of consciousness, where it can become 'deliberative cognition' (DiMaggio 1997: 269–72, see also D'Andrade 1995). As DiMaggio (1997: 271) notes, 'When sufficiently motivated, people can override programmed modes of thought to think critically and reflexively'. The moment of this shift from automatic cognition to deliberative cognition provides a window through which cognitive operations can be revealed and analysed.

Social dramas can also offer insights into cognitive processes. In the course of social dramas, the role of emotions 'can be accessed more easily … because participation in dramatic incidents also intensifies emotional involvement, generally making it more visible. Moreover, *emotional displays* serve as a central communication medium in social conflicts' (Röttger-Rössler 2009, emphasis in the original).[12]

A highly innovative methodology has been outlined by Kirill Istomin – one of the contributors to this volume – and Juan Domínguez Duque in a project proposal inspired in part by some of the work of Robert Turner (see also Domínguez Duque et al. 2010). Turner is one of the directors at the Max Planck Institute for Human Cognitive and Brain Sciences in Leipzig, Germany, and a pioneer in the fields of brain physics and functional magnetic resonance imaging (fMRI). He has conducted experiments demonstrating how cultural activities and collective representations can change the actual physical structure of the brain (see Turner and Whitehead 2008). In doing so, Turner has in effect proven the assumption of Donald Hebb, who described a mechanism of synaptic plasticity that has been explained as 'cells that fire together wire together' (Turner and Whitehead 2008: 48, see also Reyna 2002: 70, Doidge 2007: 427). This is known as 'Hebbian learning',

11 According to Lees and Bates (1990: 266), such events 'might include a change in the market price of a cash crop, or a land-reform law which alters land tenure, or a relatively large-scale migration into or out of a locality, or establishment of a quota on certain types of livestock'.

12 See also Zerilli (2006) on 'sentimental dramas'. For an introduction to the cognitive structure of emotions, see Ortony, Clore and Collins (1988).

the mechanisms of which include 'synaptic potentiation and dendritic sprouting' (Turner and Whitehead 2008: 48). In other words, repeated and persistent activity can lead to a process of cellular growth and lasting structural change within the brain that is observable using brain imaging techniques. Istomin and Domínguez Duque have designed a meticulous methodology that actually includes using such brain imaging techniques.[13]

Istomin and Domínguez Duque's project focuses on wayfinding and navigation, which are rather discrete activities and therefore more amenable to such types of brain-imaging techniques. Likewise the examples in Turner and Whitehead's paper focus on physical changes to brain structure as a result of certain repeated cultural motor activities such as playing musical instruments, which show up on fMRIs. MRIs have also been used to demonstrate the effects of meditation on the structure of the brain. But could such imaging techniques be used to study and analyse something that is more in the realm of unconscious beliefs such as, say, category-based identities and identity formation of the type I discussed earlier? Turner and Whitehead (2008: 49) suggest that 'it is at least plausible that most collective representations have some form of brain representation'. Perhaps there are measurable structural changes in the brain that occur in the processes of identity formation and repeated assertion of identity, if not in the thickening of the cortex then perhaps in some other form of 'brain representation' of collective representations (Turner and Whitehead 2008: 49–53).

As previously noted, Brubaker, Loveman and Stamatov (2004) suggest that two different ways of cognizing identity can exist simultaneously. Perhaps Turner's type of brain-imaging approach could reveal the particular regions of the brain that are stimulated by interaction with these different types of attachments and identities. Identities that form as a result of dwelling and emplacement in a given place may be more deep-seated. Perhaps they are more emotionally loaded and especially salient, which may stimulate more activity in the amygdala or other parts of the limbic system, whereas the brain activity stimulated by identities based on externally imposed categories or on the Andersonian type of 'imagined communities', in other words, on an abstract and externally constructed idea of a shared territory, may be restricted to the prefrontal cortex. However, given the relative plasticity of the brain and the malleability of cognition, it could be that these two ways of cognizing identity become so tightly associated over time that they in effect combine to form a single schema. In this case, representations of either would stimulate the same cognitive processes for both – in other words, such a combination of types of identifications could create new neural pathways. Or it could be that the more abstracted sense of identification with an imagined community becomes strong enough and deeply enough seated that it starts to stimulate the more emotional or limbic type of reaction after some time, and

13 The project is titled *Brainscapes: Investigating How Differences in Environmental Layout Influence the Neural Topography of Spatial Navigation*. The project's pilot phase was funded by the Max Planck Network for Cognitive Research (MaxNet Cognition).

could even push out an earlier deep-seated identification based on dwelling and emplacement. This is all entirely speculative, but such are the types of questions that could perhaps be investigated through a combination of anthropological methods and brain imaging techniques.

Concluding Remarks

I am not suggesting that all anthropologists should now run into the field with the as-of-yet uninvented portable fMRI cap. Nor do I propose that all anthropologists even need to explicitly engage with this matter, just as I mentioned at the beginning that we decided to avoid it in our paper on identity (Donahoe et al. 2009). Anthropology is a large field, and there is plenty of room for such cognitive approaches as well as approaches that will never come anywhere near touching it.

Anthropology is probably the social science discipline most resistant to engaging with cognitive and brain sciences, and especially with the real bogeyman, evolutionary psychology. Now, as this is an 'impulse' essay, the point of which is to generate discussion, I would like to leave with a few open questions targeted primarily at anthropologists.

What do we as anthropologists, and anthropology as a discipline, stand to gain from serious engagement with brain and cognitive sciences? And what do we stand to lose? What are the dangers?

Conversely, what do we gain by *not* engaging with these sciences, and what do we lose?

If we, as anthropologists, want to understand cognition – which is, I would argue, necessary for a better understanding of the ways in which humans, cultures and natures or environments interact and mutually constitute one another – and if we are going to speculate about cognition, using concepts that are inherently cognitive, then we should do so in a way that is consistent with the most recent developments in cognitive and brain sciences. The dogmatic insistence on nurture to the almost complete exclusion of nature that has dominated sociocultural anthropology for much of the past thirty years can only rest on an outmoded understanding of the brain as a virtual *tabula rasa* that is waiting to be filled up with and shaped by 'culture'.[14] But the human brain is a highly evolved organ that coordinates a multitude of functions occurring in specific areas of the brain; this evolved structure of the brain is the same across cultures. The understanding of the brain has steadily progressed from the *tabula rasa* metaphor to the theory of domain-general modules, then to the idea of domain-specific modules as rather rigid and inflexible evolved innate structures (as in much of evolutionary psychology), and, most recently, to the more sophisticated theory that brain structures 'arise during development as a result of the interaction of innately

14 For recent critiques of the *tabula rasa* approach, see Pinker (2002) and Ridley (2003).

sketched-out pathways and their terminating areas with repeated experience that produces Hebbian learning' (Turner and Whitehead 2008: 54, n. 4, referring to Karmiloff-Smith 1995). No doubt the field will continue to develop in ways that will make today's most cutting-edge theories look naïve and facile. Nevertheless, this last theory assumes the evolved *plasticity* of the brain and its capacity to develop and change its structure through interaction with the social, cultural and physical environments; as such, it is completely compatible with sociocultural anthropology's aims.

That is the first level of engagement – simply to be informed of current developments in brain and cognitive sciences and make sure that our speculations, assertions, and theorizing at least do not fly in the face of those findings. A next level of engagement would be to actively employ such findings in our research, such as Edwin Hutchins did in *Cognition in the Wild* (1995). A further level of engagement would be to interact directly with cognitive scientists in the development of our and their research designs, methodologies, and theorizing, as suggested by Steve Reyna in his book *Connections* (2002) and as I think Kirill Istomin and Juan Domínguez Duque's project aspires to do. In doing so we also have something important to contribute to brain and cognitive sciences through our methodological specializations that take brain and cognitive sciences out of the laboratory and can potentially bring them into the wild. With this approach I believe we can turn genetic determinism and evolutionary psychology on their heads and show that culture is now driving the evolutionary bus. That would be an important contribution that would also raise sociocultural anthropology's relevance in these debates. And finally, the more seriously we can engage with these sciences, at whatever level, the more credibly we can argue against them when necessary, to counter the misapplications and abuses that they lend themselves to. That, I think, should appeal to all anthropologists who have a vested interest in addressing what they see as abuses of these sciences. Screaming that they are absurd, reductionist, deterministic and essentialist is simply not enough.[15]

What do we stand to lose by engaging with the cognitive sciences? We run the risk of losing sight of the cultural forest for the cognitive trees, so to speak; of losing sight of our own methods, of being forced to take on certain assumptions and approaches that are perhaps irreconcilable with ours in order to be taken seriously by scientists in these fields; in other words, we run the risk of losing our very identities as anthropologists. We run the risk of being tarred by the same brush (reductionism, biological/genetic determinism) as sociobiology and evolutionary psychology. We run the risk of giving succour to the enemy, and of having our research misunderstood and misapplied by those who take it out of context to support scientifically unfounded and politically unpalatable positions such as racism. We run the risk of being ostracized by the majority of sociocultural

15 See, for example, Tim Ingold's commentary on Francisco Gil-White's article in *Current Anthropology* (2001).

anthropologists. These are real risks that we must take seriously, but they are avoidable.

What do we lose by not engaging? Well, to the degree that some anthropologists still insist on mystical explanations of the brain-body-environment matrix of interactions and reject the new findings in cognitive sciences on the basis of caricatured notions of what they do, we look and sound like ostriches with our heads in the sand, and we marginalize ourselves in important public and scientific debates. And that is truly a great loss.

Acknowledgements

This essay was first presented at the conference *Nomadic and Indigenous Spaces: Productions and Cognitions*, held in Leipzig, Germany, 3–4 February 2011. The conference organizers requested that I present an impulse paper on the theme '"Indigenous", "tribal", "nomadic": categories, their construction and their analytical and political contents'. As such, the essay was intended to be somewhat antagonistic and controversial to generate debate. I have intentionally left it that way. I would like to thank the conference organizers for the invitation; John Eidson, Kirill Istomin, Judith Miggelbrink, Benjamin Purzycki and an anonymous reviewer for their comments on an earlier draft; and Stephen Reyna for his prompt and insightful responses to my queries. While I have not incorporated all of their suggestions or addressed all of their concerns, the essay has benefited significantly from their insights. I take full responsibility for any remaining shortcomings.

References

Agrawal, A. 2005. *Environmentality: Technologies of Government and the Making of Subjects*. Durham, NC: Duke University Press.

Appadurai, A. 1988. Putting hierarchy in its place. *Cultural Anthropology*, 3(1), 36–49.

Appadurai, A. 1998. Number in the colonial imagination, in *Modernity at Large: Cultural Dimensions of Globalization*. Minneapolis, MN: University of Minnesota Press, 114–35.

Baars, B. 1986. *The Cognitive Revolution in Psychology*. New York: Guilford Press.

Banks, M. 1996. *Ethnicity: Anthropological Constructions*. London: Routledge.

Barrier, N.G. (ed.) 1981. *The Census in British India*. New Delhi: Manohar Publications.

Barth, F. 1969. Introduction, in *Ethnic Groups and Boundaries: The Social Organization of Culture Difference,* edited by F. Barth. Bergen: Universitetsforlaget, 9–38.

Barthes, R. 1996 [1982]. Inaugural Lecture, Collège de France, in *A Barthes Reader,* edited by S. Sontag. Scarborough, ON: HarperCollins Canada, Ltd, 457–78.

Basso, K. 1996. *Wisdom Sits in Places*. Albuquerque: University of New Mexico Press.

Benedict, R. 1934. *Patterns of Culture*. Boston, MA: Houghton Mifflin.

Berlin, B., Breedlove, D. and Raven, P. 1966. Folk taxonomies and biological classification. *Science*, 154, 273–5.

Berlin, B. and Kay, P. 1969. *Basic Color Terms: Their Universality and Evolution*. Berkeley, CA: University of California Press.

Bourdieu, P., Wacquant, L. and Farage, S. 1994. Rethinking the state: Genesis and structure of the bureaucratic field. *Sociological Theory*, 12(1), 1–18.

Brubaker, R., Loveman, M. and Stamatov, P. 2004. Ethnicity as cognition. *Theory and Society*, 33, 31–64.

Cheshko, S.V. 2005 [2000]. The population census: Whom to count and how? *Anthropology and Archeology of Eurasia*, 44(1), 8–24.

Cohn, B.S. 1987. The census, social structure and objectification in South Asia, in *An Anthropologist among the Historians and Other Essays*, edited by B.S. Cohn. New Delhi: Oxford University Press, 224–54.

Colby, B. 1996. Cognitive anthropology, in *Encyclopedia of Cultural Anthropology*, vol. 1, edited by D. Levinson and M. Ember. New York: Henry Holt & Co., 209–14.

Conklin, H. 1954. *The Relation of Hanunóo Culture to the Plant World*. Ph.D. dissertation, Yale University.

Conklin, H. 1980. *Folk Classification: A Topically Arranged Bibliography of Contemporary and Background References through 1971*. New Haven, CT: Yale University, Department of Anthropology.

Cornell, S. and Hartmann, D. 1998. *Ethnicity and Race: Making Identities in a Changing World*. Thousand Oaks, CA: Pine Forge Press.

Damasio, A. 2010. *Self Comes to Mind: Constructing the Conscious Brain*. New York: Pantheon.

D'Andrade, R. 1995. *The Development of Cognitive Anthropology*. New York: Cambridge University Press.

DiMaggio, P. 1997. Culture and cognition. *Annual Review of Sociology,* 23, 263–287.

Doidge, N. 2007. *The Brain that Changes itself.* New York: Viking Penguin.

Dominguez Duque, J., Turner, R., Lewis, D., and Egan, G. 2010. Neuroanthropology: A humanistic science for the study of the culture-brain nexus. *Social Cognitive and Affective Neuroscience*, 5(2), 138–47.

Donahoe, B. 2009. Situated bounded rationality: Linking institutional analysis to cognitive, processual, and phenomenological approaches in anthropology. *Max Planck Institute for Social Anthropology Working Papers*, 117 [Online]. Available at: http://www.eth.mpg.de/pubs/wps/pdf/mpi-eth-working-paper-0117.pdf [accessed: 1 November 2012].

Donahoe, B. 2011. On the creation of indigenous subjects in the Russian Federation. *Citizenship Studies*, 15(3–4), 397–417.

Donahoe, B., Eidson, J., Feyissa, D., Fuest, V., Hoehne, M., Nieswand, B., Schlee, G. and Zenker, O. 2009. The formation and mobilization of collective identities in situations of conflict and integration. *Max Planck Institute for Social Anthropology Working Papers*, 116 [Online]. Available at: http://www.eth.mpg.de/pubs/wps/pdf/mpi-eth-working-paper-0116.pdf [accessed: 1 November 2012].

Durkheim, E.. 1965 [1912]. *The Elementary Forms of Religious Life*. New York: Free Press.

Ellen, R. 2006. *The Categorical Impulse: Essays in the Anthropology of Classifying Behaviour*. New York: Berghahn Books.

Erikson, T. 1993. *Ethnicity and Nationalism*. London: Pluto Press.

Florence, M. [M. Foucault]. 1984. Foucault, in *Dictionnaire des philosophes*, 942–4 [Online]. Available at: http://foucault.info/foucault/biography.html [accessed 7 October 2011].

Gardner, H. 1987. *The Mind's New Science: A History of the Cognitive Revolution*. New York: Basic Books.

Gil-White, F. 2001. Are ethnic groups biological 'species' to the human brain? Essentialism in our cognition of some social categories. *Current Anthropology*, 42(4), 515–54.

Hacking, I. 1982. Biopower and the avalanche of printed numbers. *Humanities in Society*, 5(3–4), 279–95.

Harrison, D. 2007. *When Languages Die: The Extinction of the World's Languages and the Erosion of Human Knowledge*. Oxford: Oxford University Press.

Helling, D. 2009. *Anatomy of a 'Political Chameleon': Re-Examining Fluid Shapes and Solid Constants of Nationalism and Nation-building*. Crisis States Discussion Papers, 17. London: London School of Economics, DESTIN Development Studies Institute.

Hirschman, C. 1987. The meaning and measurement of ethnicity in Malaysia: An analysis of census classifications. *Journal of Asian Studies*, 46(3), 552–82.

Hutchins, E. 1995. *Cognition in the Wild*. Cambridge, MA: MIT Press.

Ingold, T. 2000. *The Perception of the Environment: Essays on Livelihood, Dwelling and Skill*. London: Routledge.

Jenkins, R. 1997. *Rethinking Ethnicity*. London: Sage.

Karmiloff-Smith, A. 1995. *Beyond Modularity: A Developmental Perspective on Cognitive Science*. Cambridge, MA: MIT Press.

Kertzer, D.I. and Arel, D. (eds). 2002. *Census and Identity: The Politics of Race, Ethnicity, and Language in National Censuses*. Cambridge: Cambridge University Press.

Lees, S. and Bates, D.G. 1990. The ecology of cumulative change, in *The Ecosystem Approach in Anthropology: From Concept to Practice*, edited by E.F. Moran. Ann Arbor, MI: University of Michigan Press, 247–77.

Levinson, S. 1998. Studying spatial conceptualization across cultures: Anthropology and Cognitive Science. *Ethos*, 26(1), 7–24.

Lizardo, O. 2004. The cognitive origins of Bourdieu's *habitus*. *Journal for the Theory of Social Behaviour*, 34(4), 375–401.

Ortony, A., Clore, G. and Collins, A. (eds). 1988. *The Cognitive Structure of Emotions*. Cambridge: Cambridge University Press.

Pinker, S. 2002. *The Blank Slate: The Modern Denial of Human Nature*. New York: Viking Penguin.

Porter, T. 1995. *Trust in Numbers: The Pursuit of Objectivity in Science and Public Life*. Princeton, NJ: Princeton University Press.

Quinn, N. and Straus, C. 2006. Introduction to special issue on 'The missing psychology in cultural anthropology's key words'. *Anthropological Theory*, 6(3), 267–79.

Reyna, S. 2002. *Connections: Brain, Mind, and Culture in a Social Anthropology*. London: Routledge.

Ridley, M. 2003. *The Agile Gene: How Nature Turns on Nurture*. New York: HarperCollins.

Röttger-Rössler, B. 2009. End of honor? Emotion, gender, and social change in an Indonesian society, in *Emotions as Bio-Cultural Processes*, edited by B. Röttger-Rössler and H.J. Markowitsch. New York: Springer, 317–28.

Sokolovski [Sokolovskii], S.V. 2005a. Identity politics and indigeneity construction in the Russian Census 2002. *Max Planck Institute for Social Anthropology Working Papers*, 77 [Online]. Available at: http://www.eth.mpg.de/pubs/wps/pdf/mpi-eth-working-paper-0077.pdf [accessed: 1 November 2012].

Sokolovskii, S.V. 2005b [2000]. The 2002 Census: Games according to Wittgenstein. *Anthropology and Archeology of Eurasia*, 44(1), 25–33.

Stepanov, V.V. 2005 [2001]. The All-Russia 2002 Census: Ways to measure the identity of large and small groups. *Anthropology and Archeology of Eurasia*, 44(1), 34–94.

Strauss, C. and Quinn, N. 1997. *A Cognitive Theory of Cultural Meaning*. Cambridge: Cambridge University Press.

Thornton, T. 2008. *Being and Place among the Tlingit*. Seattle: Washington University Press.

Turner, R. and Whitehead, C. 2008. How collective representations can change the structure of the brain. *Journal of Consciousness Studies*, 15(10–11), 43–57.

Urla, J. 1993. Cultural politics in an age of statistics: Numbers, nations, and the making of Basque identity. *American Ethnologist*, 20(4), 818–43.

Zerilli, F.M. 2006. Sentiments and/as property rights: Restitution and conflict in postsocialist Romania, in *Postsocialism: Politics and Emotions in Central and Eastern Europe*, edited by M. Svasek. New York: Berghahn Books, 74–94.

Chapter 8

Learning to Be Seated: Sedentarization in the Soviet Far North as a Spatial and Cognitive Enclosure

Joachim Otto Habeck

> A calm, cloudless day in July [1940]. Vivi, a small trading post consisting of five houses, nestles on the steep bank of the Lower Tunguska. ... Today, at the assembly of the Stalin collective farm in Vivi, members will take a decision on a question at the top of the agenda – the question of whether to convert to a sedentary life. [After this introduction, the author of the newspaper article changes into past tense.] ... The hunter F.P. Pankagir was the first to speak in the debate. He said: 'Under Tsarism, I never had more than six reindeer, now I own more than 30 thanks to the help of Soviet power. We no longer have to travel hundreds of kilometres to buy flour or fabric, all this is now being delivered to the workers' brigades and this has increased our income. We now have iron stoves and kerosene lamps in our tents. In earlier times, I never saw an Evenki child attending school; now all of our children learn in them. If the Soviet government, our Bolshevik Party, and our beloved Comrade Stalin tell us that we should change to a sedentary life and provide us with plenty of help to do so, then we will do it. For my part, I really want to replace life in a tent with life in a house.' ... One after the other, the members of the collective farm spoke up and expressed their love of the government, the Party, and the leader of the peoples, Comrade Stalin. And when the chairman proposed a vote in favour of transition to a sedentary life, the hands of the collective farm members all rose at once. (Anonymous, newspaper article in the 1941 special issue of *Evenkiiskaia Novaia Zhizn'*, translation: Joachim Otto Habeck)

This chapter serves to illustrate how sedentarizing nomads works in practice, more specifically, how it worked in the remote parts of the boreal forest (*taiga*) of the Soviet Union in the 1940s and 1950s. The different aspects of the Soviet state's sedentarization campaign – administrative restructuring, collectivization, and developing new industrial branches – will be illuminated through examples of official documents and responses by Siberian reindeer nomads who were affected by this policy. In particular, the focus will be on reindeer nomads' comments on how their spatial practices are subject to state-instigated change.

I shall seek to develop the concept of 'socialist land enclosure' (Fondahl 1998, see also Anderson 1992) and complement it with the idea of *cognitive enclosure*, by which I mean the very palpable consequences of sedentarization on people's perception of space and skills of movement. Here a first definition: we can speak of 'cognitive enclosure' when a human population, or a significant part of a population, no longer has access to the land they formerly lived on and therefore gradually forgets the skills, the routes, and the landmarks they knew while living on that land.[1] I will discuss the outcomes on two levels. First, I will take a macro perspective, and the scale of processes I am going to analyse – the distance of relocation – is measured in tens or even hundreds of kilometres. Then, I will deal with the consequences at the micro level of transition to sedentary life, notably, with changing practices of sitting, squatting and lolling in different types of dwellings.

It is apposite to start with a remark on the regional setting and the sources of this chapter. I draw mainly on material from Central Siberia. It is inhabited by several indigenous groups, the most widespread and numerous of which is that of the Evenkis, formerly known as the Tungus. Notwithstanding the vast area that Evenkis inhabit, and despite substantial differences in livelihood and other aspects, the Evenkis have mostly been considered as a single ethnic group by officials, travellers to the region, ethnographers, and anthropologists. The Evenkis are renowned as hunters and reindeer nomads of the taiga, and they are the protagonists of many ethnographies, in Russian (for example, Sirina 1995, Tugolukov 1985, Turov 1990, Vasilevich 1969) and English (for example, Anderson 2000, Bloch 2004, Brandišauskas 2009, Fondahl 1998, Povoroznyuk 2007, Safonova and Sántha 2011, Sirina 2006, Turov 2010). In order to interpret and contextualize the Evenki case more widely, reference will be made to other ethnic groups in the Far North of Russia, notably the Komi (Zyrians), the Sakha (Yakuts) and the Chukchi.

More narrowly, the archival sources and biographic accounts used in this chapter stem from three sub-regions within Central Siberia: the Taimyr region, where David G. Anderson worked for extended periods (Anderson 2000); northern Transbaikalia, which was described by Gail Fondahl in her 1998 monograph; and the region between the two – formerly known as the Evenki Autonomous Okrug, where I conducted archival research in 1995 (Habeck 1998). Fortunately, I can draw not only on textual but also visual sources from the same region: members of the Soviet Polar Expedition 1926–27 produced a host of intriguing photographs and drawings in the region under study, and it is thanks to Anderson and his colleagues that these resources have been retrieved from Siberian archives and made accessible online.[2]

1 I am aware that the concept of cognitive (en-)closure is used in psychology, notably by John Duncan (2010), who with this term describes the propensity of the human brain to focus on certain possibilities of structuring information, discarding the multiple other potential, but unlikely, possibilities.

2 Photographs and drawings referred to in this chapter are stored in the Krasnoiarsk Territorial Museum of Regional Studies (*Krasnoiarskii kraevoi kraevedcheskii muzei*,

The Evenki Autonomous Okrug (*Evenkiiskii avtonomnyi okrug*) existed as an administrative unit from 1931, when it was founded as part of the Soviet policy of giving indigenous groups their 'own' titular territories, to 2006, at a time when several of those ethnically defined territories were merged with other administrative units as part of Putin's policy of centralization. The territory is extremely sparsely populated (17,000 inhabitants on an area of the size of Turkey). In the past, the majority was nomadic, whereas now only the minority of inhabitants stays in the forest for long periods; however, hunting and fishing (and in specific areas also reindeer herding) continue to play a key role in people's livelihoods.

The Fate of Vivi

Vivi, its history and the events of summer 1940 reported at the beginning of this chapter have exemplary character for the Soviet Far North in general. The site of that story is located at the mouth of the river Vivi into the Lower Tunguska River, far away from any bigger settlement. Exemplary is also the fact that Vivi is uninhabited nowadays, although it acted as a somewhat important place in the first half of the twentieth century, first as a trading post (Figure 8.1) and then as the central base of a collective farm (*kolkhoz*). The trading post (*faktoriia*) was founded there in the mid-1920s,[3] when competing Soviet trade organizations and cooperatives established numerous small posts at the Lower Tunguska, some of which only operated for a couple of years, up to 1928 when the rival organizations were united under one umbrella (Tarasenkov 1930: 368–70). Evenki hunters more or less regularly visited the trading posts to exchange furs for tea, sugar and flour, ammunition, and other commodities. At such occasions they sometimes pitched their conical tents (Figure 8.2) in the vicinity of the respective trading post. Vivi was among the trading posts to survive and in the mid-1930s, when the Soviet government began to implement collectivization and sedentarization in this region, it was determined to become one of the new settlements where Evenkis were henceforth to reside.

Contrary to the enthusiastic tone of the newspaper article (which was published in a special issue of the regional newspaper but was apparently intended to impress readers in Moscow and other Soviet cities), the construction of a base for sedentary life in Vivi was not without difficulties. Having decided in June 1940 to take part in the building of the collective's central premises and to construct houses nearby

henceforward abbreviated as KKKM). They have been made available online on www.abdn.ac.uk/polarcensus/. A detailed description of the Polar Census 1926–27 has recently been published by a team of researchers led by David G. Anderson (Anderson 2011, Anderson and Campbell 2009).

3 A small number of trading posts and other small settlements with wooden houses already existed in this region in the late nineteenth century; however, the overwhelming majority of Evenkis (Tungus) did not live in them and a planned network of settlements only came into existence from the 1930s onwards.

Figure 8.1 Construction of a new building at the trading post (*faktoriia*) of Vivi, Lower Tunguska. Photograph taken by Nikolai P. Naumov probably in April 1927. (The newspaper article quoted at the beginning of this chapter was written fourteen years after the photograph was taken)

Source: www.abdn.ac.uk/polarcensus/Naumov.htm (KKKM 7930-1-05-02)

Figure 8.2 Members of the Pankagir clan in front of their conical tents. Enlarged portion of a photograph taken by Nikolai P. Naumov in early October 1926 near the mouth of the river Vivi

Source: www.abdn.ac.uk/polarcensus/Naumov.htm (KKKM 7930-1-09-01)

where their families would live, the assembly of the collective named after Stalin came together again in August the same year. They complained that the team of land surveyors, who apparently were not acquainted with the locality, had made a grave mistake:

> Pankagir, Mariia – the site is unsuitable for construction, as all buildings will be located on swampy ground. Pankagir, I.A. – the site is dirty and people will be unable to live at this place. Pankagir, A.A. – the expedition [of land surveyors] did not explain anything to the members of the collective farm, and so it happened that a swampy area was selected for construction. (GAEAO 7-1-112-11)[4]

The assembly then decided that the head of the team of surveyors was to cover the expenses for relocating the one building already erected on the wrong place.[5] The new settlement gradually came into existence, but it was very short-lived. Already in 1952, at the height of the first post-war campaign for merging small collective farms into larger ones, the members of the collective farm of Vivi were no longer supposed to live there, and the wooden houses that had been built over the previous years were to be disassembled and rebuilt elsewhere.[6] For a decade, the small settlement had functioned as the economic and administrative centre of part of the Evenki population. From now on it was officially declared 'uninhabited' (*nezhiloe* on topographical maps), and even though some Evenki hunters and travellers on the boats up and down the river occasionally stayed in the erstwhile settlement, people were no longer supposed to reside there. The Pankagir clan was relocated to larger settlements, where they mingled with members of other Evenki clans and with the rapidly increasing number of Russians and other incomers.

The above account of the short-lived history of Vivi and the photographs enable us to connect large-scale state-induced projects such as sedentarization with the names and faces of individual people. The story of Vivi sets the ground for what follows: a general outline of how the Soviet state succeeded in making nomads sedentary, in making them leave their accustomed areas and move into larger structures – larger settlements where state officials could educate, integrate, and control the indigenous population much more efficiently. Towards the end of the chapter, I will return to the photographs, discussing some elementary – in fact, corporeal – consequences of making people 'sedentary'.

---

4 *Gosudarstvennyi arkhiv Evenkiiskogo avtonomnogo okruga* (State Archive of the Evenki Autonomous Okrug), henceforward abbreviated as GAEAO. See list of references.

5 Stated in the same document (GAEAO, 7-1-112-11). Errors of this sort were made elsewhere, too: in some cases, surveyors and officials did not pay attention to local permafrost dynamics (Habeck 1998: 44–5).

6 Here Vivi is exceptional: the collective farm was not simply merged with others into a larger unit, but split and assigned to two different larger collective farms: to Uchami further downstream and Tura further upstream (personal communication with Altynai Pankagir, July 1995).

A Brief Timeline of the Soviet Modernization Project in the Far North

The political principles of the early Soviet approaches towards ethnic minorities in Siberia, their further formulation by the Communist Party and several state authorities, and the institutions to implement these measures have been described elsewhere (for example, Forsyth 1992, Slezkine 1994, Uvachan 1984). Here the focus lies on how this policy was implemented and what it meant for the sphere of activity and mobility of the indigenous population.

The summer of 1940 was neither the beginning nor the end of the sedentarization campaign, although it does mark an important interim stage. Soviet plans to make the whole nomadic population sedentary had been pursued since 1928, with an early focus on the steppes of Central Asia. Only later, in 1935, did serious attempts at sedentarization start in the taiga and tundra.[7] When in September 1941 the Soviet Union was attacked by Germany, the collectivization and sedentarization campaign had to be put aside, to be resumed in the late 1940s.

Figure 8.3 'View of the centre of Kislokan settlement, nomads' council (*kochevoi sovet*) of Amo'

Source: Drawing in the 1941 special issue of *Evenkiiskaia Novaia Zhizn'*: 15, caption translated by Joachim Otto Habeck

The findings of my archival research in the Evenki Autonomous Okrug (Habeck 1998) articulate well with evidence from the adjacent Taimyr region (Anderson 2000) and Northern Transbaikalia (Fondahl 1998). This allows me to sketch a

7 The striking exception in this timeline is the case of sedentarization among the Tofa in southern Siberia, completed in a very short period (1928–32) and celebrated as an outstanding story of success of the Soviet collectivization and modernization policy (Donahoe 2006: 93–5, Slezkine 1994: 279).

general timeline of sedentarization in the boreal forest zone of the Soviet Union. The most decisive steps are listed here in logical and chronological order, although this does not mean that the process always went smoothly and orderly – in fact, there are many accounts of the overhasty and slapdash character of these operations:

1. Since exact information on the indigenous inhabitants and remote Northern territories was very limited, a census appeared to be the precondition for gaining actual power and taking any further action. The Soviet census of 1927 was accompanied by a very thorough – and in many ways ethnographic – investigation of the Far North and Far East of the Soviet Union, which came to be known as the Polar Census (*Pripoliarnaia perepis'*) of 1926 and 1927 (Anderson 2011). It is thanks to the census takers that we are able to draw on rich and detailed visual resources of life in the taiga before collectivization, despite the fact that the census itself paved the way for the subsequent state-driven social engineering.
2. In nearly twenty locations across the Far North and Far East of the Soviet Union, so-called cultural bases (*kul'tbazy*) were established. They served as initial points for coming into contact with and building up trust among indigenous inhabitants. They were also designed to demonstrate the advantages of a sedentary and 'cultured' life. One of the earliest cultural bases was founded in 1927 (Habeck 1998: 28–30, 36–7), some two hundred kilometres upstream from Vivi. The cultural base then further developed into Tura, the capital of the Evenki Autonomous Okrug, eventually becoming the place of residence of Pankagir and many other Evenki families.
3. On the basis of processed census data and the observations of the first cohort of officials, planners identified 'gravitation centres' (*tsentry tiagoteniia*) of the nomadic population. These could be, as one may expect, trading posts; but in several cases, planners determined the long-standing sites for clan meetings (*suglan*) as 'gravitation centres' and thus as potential centres of future collective farms and rural settlements.
4. Collectivization, violent and unsuccessful in the first stage of the early 1930s, was subsequently conducted at a slower pace. Rather than fully-fledged collective farms, Evenki households were to join Elementary Production Associations (*prosteishie proizvodstvennye ob"edineniia*, abbreviated as PPO). This temporary concession to the 'less developed' Northern peoples began in 1933 and ended shortly after the end of the war, when all PPO were turned into collective farms (*kolkhozy*).
5. The initial land allocation (*pervonachal'noe zemleustroistvo*), carried out between 1936 and 1939, assigned discrete territories on which the PPO and subsequent collective farms were to operate. Despite not having binding force in practice in the first years, the delineation of collective-farm borders reduced the flexibility of earlier land-use arrangements and the multiplicity of entitlements.

6. As a further step, internal land allocations (*vnutrennee zemleustroistvo*) were conducted from 1939 to 1941 to determine, among other things, the location for the central base of the nascent collective farm. A seemingly easy task, this measure sometimes resulted in serious problems, as in the case of Vivi.
7. The construction of wooden houses for PPO/collective farm members already started in 1936. Figure 8.3 shows an idealized image of this activity: the emergence of neat houses on the banks of the Lower Tunguska. The drawing is taken from the same 1941 newspaper that features the article about Vivi. Most houses, however, were built after the war. In the Evenki Autonomous Okrug, as probably in other regions, too, loans were provided for this purpose, yet there are reports that these loans were granted with considerable delay and were generally uncommon in the first years (GAEAO 7-1-95-1). Officials encouraged Evenki men to participate in the construction of their new homes but, more often than not, Russians and other incomers took the lead in this task. The recipients of these favours, indigenous women and men, were not always willing or able to become accustomed to their new abode. Examples of these difficulties are given below.
8. The sedentarization policy logically required establishing paid jobs. In fact, the policy was intended as a means to use the indigenous people's (in particular, women's) workforce more efficiently. While part of the now officially sedentary population still held jobs as hunters and herders and thus stayed in the taiga for extended periods, the other adult household members were expected to take up jobs in newly created industries such as fur farms. A few PPO already tried to establish fur farms in 1936. However, the peak of constructing fur farms seems to have occurred between 1947 and 1955.
9. Relatively soon, Soviet officials effected mergers of collective farms. In the small units (PPO, collective farms), people knew each other and were in fact related. In the logic of the Soviet modernization project, agricultural enterprises were to become independent of supposedly backward kin-based arrangements. Occasional mergers were already induced in the 1940s, but the first wave of mergers can be clearly associated with the early 1950s. This first wave also affected Vivi (closed in 1952).
10. The period 1964 to 1970 saw a second wave of mergers, when collective farms were amalgamated into state farms. In some cases the former collective farms were turned into branches (*otdeleniia*) of the respective state farm and the settlement of the branch continued to exist, in other cases the settlements were closed and inhabitants were relocated (again).[8] Simultaneously, hunting

8 Retrospectively, the relocations of the late 1960s appear to be particularly poorly prepared projects: being forced to leave their homes, relocated families arrived at their new place of residence only to find out that houses or flats were not available for them. See Anderson (2000: 57–9) for a Taimyr example, Allemann (2010: 88–93) for a very similar case on Kola Peninsula. He describes how many men suffered from not getting a (useful)

was now defined as a profession and came to be territorially separated from herding and other activities: in many parts of the taiga zone, state hunting enterprises (*zveropromkhozy*) were founded and hunting plots were subsequently assigned to individual hunters (see below); simultaneously, reindeer herding was made 'rational', too, with only a small cohort of men to work as herders. Generally speaking, the state authorities sought to introduce industrial methods of production into agriculture, hunting and herding.

11. As a result of the above steps, hunting and herding were pursued by few professionals whereas the rest of the family stayed in the village. This came to be known as the transition from 'nomadism as a way of life' (*bytovoe kochevanie*) to 'productive nomadism' (*proizvodstvennoe kochevanie*).

Gail Fondahl, in her study of Evenki land use and land reform in south eastern Siberia, describes the process of 'Socialist land enclosure' (1998: 73–4, see also Anderson 1992). The term denotes the process of splitting up collective hunting grounds in the taiga (i.e., common pool resources) into hunting lots to be used by individual professional hunters (thus creating individual entitlements to hunting). This process mainly took place in the 1970s and 1980s. I would argue, however, that the term can be used more generally, to include the whole process of land formation (*zemleustroistvo*) and the official delineation of territories to be used by specific collectives or individuals, from the 1920s to the 1980s, as explained above.

Mobility Regimes: Old and New Modes of Travelling

As a consequence of the Soviet modernization project in the Far North, many people who had previously migrated over smaller or greater distances across the boreal forests, started to live in villages for the most part of the year. Many women got jobs in the village administration, post office, shop, fur farm, etc.[9] Children were obliged to attend school from 1 September to early June. The long summer holiday allowed the children and their mothers to join those family members – mainly men – who stayed in the forest, for about three months per year. Thus, on the one hand, families were split; on the other hand, a sizeable number of people managed to keep a temporary link to the forest. However, this link was fragile. When in the 1990s, helicopter transport between villages and the forest ceased to operate, it was completely severed. People were stuck in the village (Campbell 2003: 115, Vitebsky 2005: 77). Transportation by reindeer had gradually been abandoned in the 1960s, and now the villagers had no reindeer to

job in their new environment, being effectively unemployed, and how the ensuing lack of motivation was socially branded as *tuneiadstvo* (idleness, parasitism).

9 Below, I will elaborate on the increased mobility that came along with state-subsidized vacations and bonus journeys.

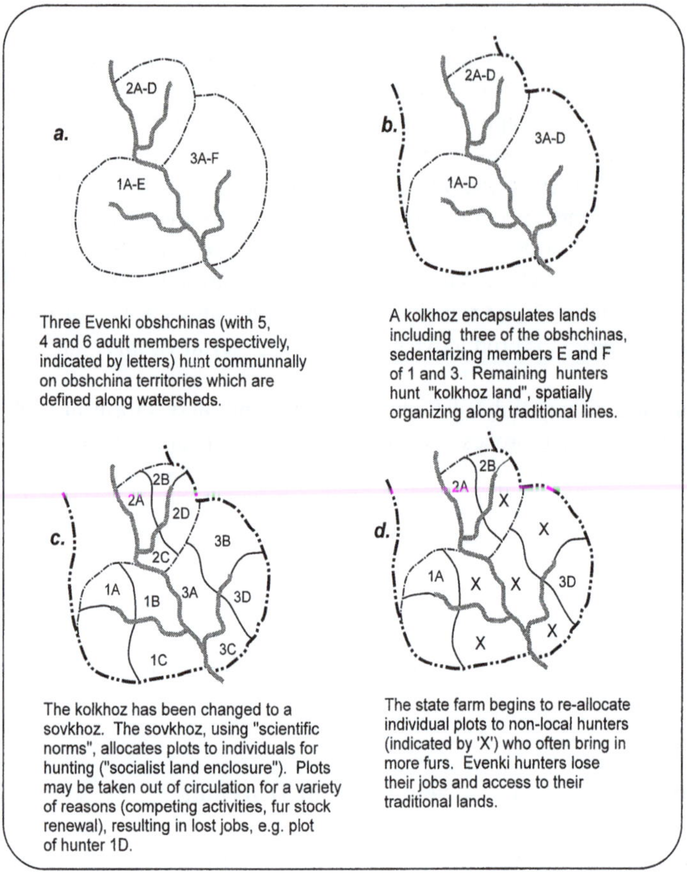

Figure 8.4 Gail Fondahl's (1998: 73) schematic representation of land tenure changes (concomitant with the process of sedentarization further north that I reconstructed from interviews and archival material. The word *obshchina* refers to a group of households, usually connected by kinship)

Source: Fondahl, Gail A., Gaining Ground?: Evenkis, Land, and Reform in Southeastern Siberia (part of the Cultural Survival Studies in Ethnicity and Change Series), 1st edition, © 1998, p. 73. Reprinted by permission of Pearson Education, Inc., Upper Saddle River, NJ

travel to the forest. And even if they had reindeer, few would remember how to actually handle the animals and how to travel through the forest.

On an even larger scale, but also in the Evenki setting, David G. Anderson (2000: 131) points to the consequences of the diminishing personal space of action. He recollects the words of a very old woman he interviewed in the early 1990s in the village of Khantaika, not far from the lower Yenisei (Map 8.1). Since this woman did not speak Russian, Anderson made use of a local translator:

Map 8.1 Central and East Siberia

During my last visit I asked her to describe the places where she travelled as a child. She spoke at great length about caravans and treacherous mountain passes. As asides, she mentioned the trading-posts of Tura and Turukhansk, and the northern outpost of Volochanka at the centre of the Khatanga Way. At these place-names my translator would smile with embarrassment and apologize, 'You have to forgive the old woman – she must be senile – it is impossible for her to have seen these places.' As the crow flies, the distances between those three points are measured in thousands of kilometers and are separated by the sharp ravines of the Putoran plateau. While it is true that commuting between these points would seem impossible to a villager dependent upon the monopolistic and inefficient schedules of the civic aviation authority, the old woman's stories do resonate with both the stories and practice of present-day *tundroviki* [i.e., reindeer herders and hunters] and the archival record. (Anderson 2000: 131, compare Ingold 2011: 149)

Anderson (2000: 131, 135) also reports that the head of the team of reindeer herders that he migrated with was not willing to specify the regional origins of his clan and simply said, 'The old people travelled everywhere', and that Evenki lands were 'everywhere'. To be sure, this reindeer herder's refusal to provide any spatial specification may have had a political slant. By claiming that Evenki lands were 'everywhere', he avoids getting bogged down in potential territorial disputes with outsiders, be they bureaucrats or anthropologists. However, taken at face value, the statement reminds us that Evenkis were once engaged in enormously extensive trading networks. Long-distance trade was carried out by households with larger herds of deer (Vasilevich 1969: 180). In Russian, these long-distance travellers were known as *kaëry*, and this word also came to designate the professional travellers that commuted with their reindeer-sledge caravans between administrative centres and remote collective farms, up to the 1960s (see Amel'kin 1995). Geologists, miners, and others who sought to open up the mineral resources of the vast expanses, also crucially depended on the reindeer-sledge caravans to reach potential deposits (Campbell 2003: 112).

Clearly not all Evenkis covered such large distances as those described by the old woman. The ambit of hunters was usually smaller in size, but trips over distances of several hundreds of kilometres occurred on a regular basis, for example for the annual delivery of furs at a trading post or the headquarters of the collective farm in the newly-erected villages. In addition to using the reindeer sledge as a means of transportation in the winter, Evenki hunters walked[10] in the snow-free period, used skis when suitable, and rode reindeer in the forest zone throughout the year.

After the end of the 1960s, the regular trips between the forest and village were conducted by helicopter instead of reindeer. Small airplanes were used for regular flights between the villages and the *okrug* centre. From the 1970s onwards, snowmobiles (skidoos, *burany*) appeared in many parts of the taiga and tundra, enabling hunters and herders to travel over short and mid-range distances (Stammler, this volume). In general, both the art of riding reindeer and the skill of steering a reindeer sledge gradually fell into oblivion, with only few experts still practicing this mode of travelling.

Tatiana Argounova-Low (2012) provides yet another striking example from the north of Central Siberia, describing how the close ties between two Sakha (Yakut) communities were severed by administrative measures. These two communities are Olenek in the Republic of Sakha (the erstwhile Yakut Autonomous Socialist

10 Walking as a mode of travelling does not receive particular attention in this contribution. Studies on Evenki specificities of travelling on foot were recently provided by Brandišauskas (2009, ch. 8) and Safonova and Sántha (2011). The latter describe walking as the most important form of companionship among Evenkis, and an activity that resonates with culturally specific forms of thinking. Long-distance trips on foot are also mentioned in the recent works of Argounova-Low (2012: 203) and Campbell (not dated).

Soviet Republic) and Yessei, which after a controversial decision in 1930 was assigned to the Evenki Autonomous Okrug (see Anderson 2000: 155–7):

> One salient story, often narrated in both Yessei and Olenek, revolves around *podriad*, so common in the past and almost not practised nowadays. *Podriad*, a long caravan of haulage reindeer, was used to bring goods from trading posts (*faktoriia*) until planes and cargo lorries partially replaced the need to travel to the hubs and the significance of roads has lessened. As one informant … described: 'My parents used to work on *podriad* to Delinde [Zhilinda], Sukhana, Saskylaakh … When the sovkhoz … was established, *podriad* disappeared. From 1961 most of the goods were delivered by planes' (Field notes, 2004). Yet until 1961 whole families were employed in *podriad* to transport food and goods. One particular journey to the farthest *faktoriia* would take 30 days. (Argounova-Low 2012: 196)

Previously connected by the frequent exchange of goods and also by persistent marriage patterns, the two communities found themselves divided by an administrative border. This separation was a gradual process. 'The border was strengthened in the 1950s when communications and regular trips by reindeer between the two villages were prohibited by the administration on the pretext of quarantine. … Later still, the irregular air connections that had started in the late 1950s were stopped altogether' (Argounova-Low 2012: 193). Nowadays, travelling from Yessei to Olenek necessitates a large detour, via Tura, Krasnoiarsk, and Yakutsk. Depending on the weather and availability of flights and seats, this journey may take as long as the journey by reindeer sledge (three weeks). From this example we may conclude that, like in other parts of the Arctic, the delineation of administrative borders often entails the channelling of movement, turning earlier multiple and multidirectional network-like arrangements into unilinear tree-like ones, strictly following the hierarchy of administrative centres from lower to upper levels and back to lower ones.[11]

These changes in the frequency, range, direction and technical means of people's movements were of such obligatory and encompassing character that they created fundamentally new mobility regimes. Craig Campbell (2003), who also conducted field research among Evenkis, describes this process as the introduction of the Soviet system of mechanized travel. This channelling of movement was particularly powerful in a context where each citizen was expected to have one single, clearly defined point of residence and 'belonging' in administrative terms (see Anderson 2000: 156). Moreover, the means of transportation, along with

11 One could add numerous examples more on how former 'north-north' connections were severed and 'north-south' connections strengthened, so that travellers to a neighbouring administrative unit first have to go south, making their way to the infrastructural baseline of Siberia – the chain of cities along the Transsiberian Railway – or even to Moscow to then travel northward and eastward again.

subsidized fuel, were increasingly provided by the state, turning knowledgeable way-finders into dependent passengers. Aviation, though fast, is vulnerable inasmuch as it depends on the flawless operation of a complex technological system (Urry 2007: 53–4, 139–40, see also Safonova and Sántha 2011: 94).

Campbell's study discusses not only the disastrous disruption of the Soviet network of transportation in the early and mid-1990s, but also mentions a fact that has thus far received little attention by social scientists writing about Siberia: as employees of state farms, hunters and herders were entitled to social benefits such as subsidized holiday trips to the Black Sea, the Caucasian spas, to Central Asia, Leningrad, Moscow, etc. Some of them also went on incentive trips, granted by the authorities in recognition of individual employees' outstanding work performance (over-fulfilment of production plans). As a consequence, some, though not all, Evenki families experienced a new and distinct form of mobility that they had hitherto not known. There are occasional reports of Siberian reindeer herders going to Prague (Vitebsky 2005: 45) or Montreal, Quebec (Campbell 2003: 111) in Soviet times. Moreover, army service led young men to distant parts of the Soviet Union or to the Soviet forces in Germany. What is remarkable about such trips in the context of the transition from a nomadic to sedentary life is that state-facilitated travel programmes and infrastructure enabled part of the population to experience new destinations, means of transportation, purposes of travel, and also new forms of sociability and company in transit (for example, when travelling by train). These new forms of mobility were created and sustained throughout the 1970s and 1980s by state enterprises, the centralized trade union (*Profsoiuz*), etc. but they could not be upheld after 1991. The collapse of aviation and the dearth of fuel left many Northern communities in the lurch. In the 2000s, the situation gradually improved, and yet for many Evenkis in the remoter settlements, mobility is nowadays much more restricted than it was in late Soviet times.

New forms of mobility also require new knowledge and skills. Learning to board a helicopter while the engine is running brings about its own challenges. Driving a car demands very specific skills, and one should be able to interpret the road signs. Public transport (for example, the Moscow metro) requires not only a general literacy and capability to read abstract maps, but also a certain amount of local knowledge. Long-distance travel by train in the Soviet Union as well as in contemporary Russia is a striking example of how passengers have to learn basic rules of conduct in the railway carriage, get used to specific technical equipment (the water tap, the boiler, etc.), find out which berths are most comfortable and which stations en route are most suitable for buying food. In sum, each mode of travelling and each means of transportation requires certain skills and knowledge, the learning of which is more or less difficult (see Aporta, Istomin, Stammler, this volume):

> For a reindeer herder or a hunter in both Yessei and Olenek, knowledge of roads differs from knowledge that exists in codified form (maps, road signs and highway road codes). This type of knowledge does not exist in a written form; instead it consists of knowledge of direction, expertise of road surfaces, including the ability

> to distinguish the quality of snow, and seasonal variations of road conditions as well as knowledge of migration routes. (Argounova-Low 2012: 203)

From my colleagues Kirill Istomin and Mark Dwyer (forthcoming), I have learnt a few things about differences in navigational skills between reindeer herders in the tundra and helicopter pilots. Depending on the means of transportation, people employ different methods of dead-reckoning, triangulation, and memorizing landmarks. If steering a helicopter is a rather complex thing to do, handling a reindeer seems even more complex, simply because one has to account for the animal's preferences and predict its behaviour. Making it from A to B is a task that can only be accomplished in cooperation between sentient humans and sentient animals. Since the skill and knowledge regarding the cooperation between reindeer and driver is particularly delicate, it is especially difficult to be restored once it is no longer regularly applied and is sinking into oblivion.

To sum up my argument to this point: *Cognitive enclosure* is the outcome of relocation, displacement or another politically caused process that makes people leave the environment they are accustomed to. As a result of cognitive enclosure, people lose the capability of making a living in a certain environment and successfully navigating within that environment over a longer period of time without technical devices. The phenomenon I describe here is clearly a collective one, though the process of forgetting occurs individually. The routine of learning – be it through own first-hand experience or by being shown by others – is interrupted. And even if members of the community obtain the opportunity to explore new places, learn new modes of travelling, and become acquainted with new ways of living, the fact that they have been moved out from a habitual setting nonetheless has a marked influence on their spatial cognition. The above story of the old woman who travelled vast distances and the young one who considered this to be impossible ('she must be senile') exemplifies a reduction of ambit that is not simply physical but also mental: for members of the younger generations, the mere idea of such travel is now beyond imagination.

Lolling, Squatting and Sitting

Thus far, my focus has been on relocation and pertinent changes in spatial cognition. These can be summarized as outcomes of sedentarization on the macro level. There is yet another aspect of cognitive enclosure I would like to elaborate on. The scale of this phenomenon can be measured in metres or even centimetres. Therefore I will discuss it under the heading of micro level. I mean the very elementary forms of movements and postures, those that usually go completely unreflected, unless they become impossible owing to some kind of impediment. In such moments of habits being disrupted, bodily movement requires reflection.

Notwithstanding a wealth of studies that fall under the rubric 'the Anthropology of the Body', postural habits have received little attention in anthropological writing since the days of Gordon Hewes (1955, 1957, 1966). Hewes produced a cross-cultural analysis of postural habits (notably sitting, squatting, and kneeling) on the basis of photographs and other visual or written illustrations. Interestingly, he noted that 'North-central Siberian cultures are not represented because of a dearth of illustrations in the small number of sources on that area available to me' (Hewes 1955: 236). The photographs and drawings of the Polar Census Expedition 1926–27 can be retrospectively used to fill this gap, even though there are technical and aesthetic aspects of photography in this period (as well as later periods) that create a certain degree of artificiality in the depiction.[12] Hewes, too, is aware of the artificiality introduced by people's posing for the picture: 'one must assume that it is possible to distinguish between postures imposed upon the subjects by the photographer, and those which are habitual' (1955: 234). On the basis of his data analysis, Hewes (1957: 128) concludes: 'Styles of house construction have their influence [on posture] (…). Terrain and vegetation may determine sitting or standing habits out-of-doors. Herdsmen in regions of tall grass have to stand to watch their stock, while in a short-grass or tundra area the herder may sit or squat'. This statement, albeit of an overly generalizing sort, points to the significance of posture in different dwellings and environments. It is fair to presume that long-term changes in residence (dwelling) or natural environment are likely to bring about changes in postural habits. However, I have not found any attempt to describe the changes in postural habits and locomotion that are effected by the transition from nomadic to sedentary life.[13]

A telling piece of information for the discussion of postural changes stems from an article by Soviet ethnographer Mikhail A. Sergeev, published in 1948, the time when the state-induced sedentarization project was in full swing in this part of Siberia:

> Real hunters, such as Tungus or Lamuts, cannot imagine how one can possibly live at one and the same place continuously. The Tungus at the Sym and Ket [rivers] did not know where to take firewood from, once they had taken all the shrubs around the house and not moved on in the taiga, [they did not know] where to source water in the winter, once a larger number of people had settled

12 The census takers used glass plate cameras. For the photograph to be sharp, the individuals had to avoid any motion for five to ten seconds. I thank David G. Anderson for reminding me of this point.

13 Pierre Bourdieu's concept of bodily hexis characterizes posture as an acquired yet persistent quality that reproduces cultural differences and class distinctions: 'Bodily hexis is political mythology realized, em-bodied, turned into a permanent disposition, a durable manner of standing, speaking, and thereby of feeling and thinking … Principles em-bodied in this way are placed beyond the grasp of consciousness, and hence cannot be touched by voluntary, deliberate transformation, cannot even be made explicit' (Bourdieu 1977: 93–4, quoted by Starrett 1995: 953). Though the concept of bodily hexis may be fruitful for a further investigation of this topic, the emphasis on permanence does not help to explain the (involuntary) transformation of postural habits and locomotion.

> down in a single place of permanent residence. After the nomads had moved into 'real' houses, quite often they deserted them and returned to their 'old state'. They were confused by the space being surrounded by walls, the size of the new dwelling, and the absence of the habitual camp fire. It was hard to get accustomed to not spending the whole everyday life on the floor, and to using tables, stools, and beds [instead] (...). The Lamuts of the Okhotsk area hewed a hole in the [wooden] floor of their new houses to set up a camp fire, and they made a hole [in the ceiling] for the smoke [to go out]. The Tungus of Tokma complained about being sick in their houses, fetching a cold [now and then]. Their wives, who were used to always sitting at one and the same place near the campfire and having everything at their fingertips, complained: 'It's bad in the house; it's dark, one has to walk'. (Sergeev 1948: 68–9, translated by Joachim Otto Habeck)[14]

At first glance, the statement that 'one has to walk' is counter-intuitive. After all, we suppose that *nomads are on the move*, and sedentary people literally 'are seated' in their houses. Why is it that these women complained that one has to walk? They said so because their domestic space of action was rather small while they were still living regularly in the tent.

In order to make this clear, I shall explain a photograph (Figure 8.5) from my more recent fieldwork among Komi reindeer herders (Habeck 2005, 2006). To show this picture may seem far-fetched, as the Komi live more than two thousand kilometres further to the west. However, they also use conical tents and migrate over long distances. The image helps us understand how people sit or squat or loll in the confined space of the tent. A young herder is cutting the hair of his relative inside the tent. Both boys sit on the planks that make up the floor in the central part of the tent. A stool is not needed, even though available: a folding stool can be seen in the background. It has been whacked into the back of the tent, the floor of which is covered by reindeer skins. Pillows and blankets are stored in the back of the tent too. They are used as bedding during the night. During the day, one can loll on that part of the floor and lean on the blankets and pillows.

Most of the domestic tasks that women are supposed to fulfil can be done within a radius of a few metres. Bodily movements in the tent clearly differ from those in a flat or house. This is even more markedly the case with Evenki conical tents than with Komi ones, because the Evenki tents used to be smaller in height and diameter. Entering the tent, making one's way through the conical tent, and eating from a table with tiny legs, all create slight yet noticeable difficulties to the

14 Sergeev's observations were published in a similar wording in a collective work of Soviet ethnographers (Levin and Potapov 1956: 566–7, English translation 1964: 508–9). A similar account from a different context (Finnish Lapland, 1936) is given by Mazzullo (2005: 189): a Sámi woman recalls the difficulties of getting used to living in the house: 'she had to learn to do everything standing on her feet, and she found that very tiring and uncomfortable'.

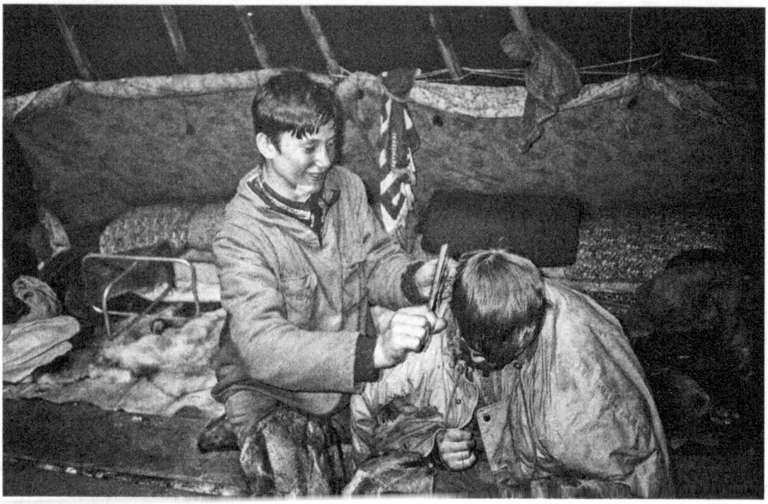

Figure 8.5 Two young Komi reindeer herders inside the tent

Source: Photograph by Joachim Otto Habeck, taken near river Sandivei, Nenets Autonomous Okrug, 4 May 1999

unaccustomed outsider who for the first time visits a Komi reindeer herders' tent. For my part, I have to say that I always found it difficult to sit in a cross-legged position, while many of the herders I lived with seemed to be at ease with it.

Looking at the archived drawings and photographs (Figure 8.6 and 8.7) that the census takers produced in 1926 and 1927 on the banks of the Tunguska and nearby rivers, we can identify many different ways of sitting and squatting in the absence of chairs. As examples I present a photograph (taken by N.V. Sushilin) and a drawing (made by A.P. Lekarenko). Comparing the two forms of depiction, the photograph seems to have more empirical validity than the drawing; however, the drawing is not influenced by the technical and aesthetic tendency of posing for the picture, mentioned above. The photograph (Figure 8.6) was probably made with the intention of minimizing the effect of posing, trying to show people as they usually behave. It depicts ten individuals, several of whom seem to pay no attention to the act of photographing. Some people are sitting on the ground (one on a felled tree) around a campfire and kitchen utensils, partly in a cross-legged position. The photograph also illustrates the relatively small size of the tent. Figure 8.7 illustrates a woman's posture and operational space inside such a tent: cooking, sewing, and various other tasks are being carried out with an upright back and legs stretched out on the floor. Necessary items are ready at hand. Since the tent is not partitioned by walls, the tools, bags, pots, etc. are usually visible and can be quickly fetched. Objects are assigned to different zones of the tent, but not yet *enclosed* in separate rooms.

To return to the women's complaint that 'one has to walk': What the women complained about is the degree of compartmentalization of their everyday sphere

Figure 8.6 An Evenki group sitting in front of a conical tent. Enlarged portion of a photograph taken by Nil V. Sushilin, 25 or 26 June 1927, near Vanavara

Source: http://www.abdn.ac.uk/polarcensus/Girshfeld.htm (KKKM 7930-1-03-08)

Figure 8.7 Picture drawn by Andrei P. Lekarenko (not dated, probably late 1920s). The caption of the drawing runs: 'In a chum. Woman at work'

Source: http://www.abdn.ac.uk/polarcensus/Lekarenko.htm (KKKM 7930-1-08-16)

of activity. They have to walk to pursue their daily tasks. Things are out of place, dispersed over the different parts of the house. For sure, many everyday routines are also spatially separate in and around the nomadic tent. Women walk around collecting firewood, berries, and water; in some cases they look after dogs or reindeer that stay in the proximity of the tent. Nonetheless I would argue that the transition from life in a tent to life in a house brings about considerable changes in the elementary forms of bodily movements, and once these changes occur, they are likely to be reflected by the individuals undergoing them. One has to learn how to live in a house. The examples I presented in this chapter elicit that this learning was a very conscious and at times even painful process.

'Chair-sitting and furniture, possibly the chief distinguishing postural attributes of Western civilization, go hand in hand … It is true that a rock, ledge, log or house platform may serve as a bench, but the fact is that people who lack special furniture for the purpose seldom sit the way we do' (Hewes 1957: 124). Thus furniture has an influence on the postures people adopt and the ways they move through the dwelling. Summarizing my thoughts about changes in bodily movements and perception of domestic space, I will close with a photograph depicting the eventual outcome of sedentarization, as envisaged by Soviet politicians, planners, and ethnographers in the mid-twentieth century (Figure 8.8). It shows a Chukchi (north-east Siberian) couple enjoying life in their newly-built and neatly decorated home. The photograph serves to demonstrate the achievement of the indigenous population of the Soviet Far North: the couple sits on chairs or stools, their arms rest on the table, which is placed right in front of the window, making the room bright and exposing the cleanliness of the dwelling. The Russian-style house was considered the most desirable type of dwelling in the 1950s, as was the flat in the Soviet multi-storey building, with central heating and running water, in the 1980s. These rooms are made for sitting, not for squatting.

Conclusion

In this chapter I have sought to demonstrate how the bureaucratic compartmentalization of nomadic space comes along with changes in skills and, importantly, the very perception of space. For these forms of compartmentalization of the everyday world, I use the term *cognitive enclosure*. I have exemplified this with different processes of learning and forgetting: losing familiarity with certain landscapes and getting used to new, more confined territories; unlearning certain modes of travelling and becoming physically and mentally dependent on new ones; leaving the tent with its specific spatial layout and learning to live in a house. These different processes, while being at work at several scales of movement, all result from a single, state-induced project carrying the label 'sedentarization'. Sedentarization was not only pursued by the Soviet state but also in Finland (see, for example, Mazzullo 2005), Canada (Dawson 2003a) and elsewhere. What is particular about the Soviet experience is the specific version of the socialist

Figure 8.8 Chukchi couple sitting at the kitchen table. The original caption runs: 'In a new house. Anadyr' raion. Collective Farm [named after] I.V. Stalin'

Source: Levin and Potapov 1956: 927, also published in Levin and Potapov 1964: 825

modernization project, with its emphasis on wholesale collectivization and the relatively uniform implementation of this project throughout the Soviet Far North.

Clearly, there are many additional aspects beyond the limited scope of this paper, for example: the issue of adapting new dwellings in ways that allow continuing habitual practices of food-processing, socializing, sleeping, etc. (Dawson 2003b) or the question of bodily movement and confinement in different means of transportation (Simonova 2007 speaks about the 'nomadic space' of carriages and trains on the Transsiberian Railway, seemingly contradicting my argument). Detailed anthropological studies are underway in Evenki communities on changing practices of travelling and interacting with reindeer (personal communications with D.G. Anderson, D. Brandišauskas and E. Landerer, autumn 2012). What I hope to have contributed to, however, is a clearer understanding of how sedentarization of nomads changes not only the targeted individuals' ambit of actions but also their bodies and perceptions.

Acknowledgements

I thank Kirill Istomin, Peter Koch, Nuccio Mazzullo, Tatiana Safonova, and István Sántha for their comments on earlier drafts of this chapter. David G. Anderson has provided many valuable suggestions and references, for which I am particularly grateful. He, Craig Campbell, Nikolai V. Martynovich from the Krasnoiarsk Territorial Museum of Regional Studies, and many other colleagues deserve special gratitude for the enormous task of making the photos and drawings of the Polar Census accessible to the public. My sincere thanks go to Altynai Pankagir and my other interlocutors in Tura (erstwhile Evenki Autonomous Okrug) and the reindeer-herding families of Ust'-Usa and Novibozh (Komi Republic), who kindly invited me to share their space of everyday life.

References

Archival sources quoted in the text

GAEAO 7-1-95-1. Gosudarstvennyi arkhiv Evenkiiskogo avtonomnogo okruga, fond 7, opis' 1, ed. khr. 95, list 1. 'Otchet ob osedanii kochevnikov po Evenkiiskomu natsional'nomu okrugu za period s 1 [ianviara] 1940 po 1 [sentiabria] 1940' [State Archive of the Evenki Autonomous Okrug, 'Report on the sedentarization of nomads in the Evenki Autonomous Okrug in the period 1 January 1940 to 1 September 1940']

GAEAO 7-1-112-11. Gosudarstvennyi arkhiv Evenkiiskogo avtonomnogo okruga, fond 7, opis' 1, ed. khr. 112, list 11. 'Protokol obshchego sobraniia chlenov Vivinskogo PPO "im. Stalina", faktoriia Vi-vi, ot 9 [avgusta] 1940 goda' [State Archive of the Evenki Autonomous Okrug, 'Minutes of the general assembly of members of the Elementary Production Association of Vivi named after "Stalin", Vivi trading post, 9 August 1940']

KKKM document numbers can be retrieved on www.abdn.ac.uk/polarcensus/

Literature

Allemann, L. 2010. *Die Samen der Kola-Halbinsel: Über das Leben einer ethnischen Minderheit in der Sowjetunion* [The Sami of the Kola Peninsula: On the life of an ethnic minority in the Soviet Union]. Frankfurt am Main: Peter Lang.

Amel'kin, A.A. 1995. *Narty* [Sledges]. Tura: Evenkiiskii okruzhnoi institut usovershenstvovaniia uchitelei.

Anderson D.G. 1992. 'Property rights and civil society in Siberia: An examination of the social movements of the Zabaikal'skie Evenki'. *Praxis International*, 12(1), 83–105.

Anderson, D.G. 2000. *Identity and Ecology in Arctic Siberia: The Number One Reindeer Brigade*. Oxford: Oxford University Press.

Anderson, D.G. 2011. The Polar Census and the architecture of enumeration, in *The 1926/27 Soviet Polar Census Expeditions*, edited by D.G. Anderson. New York: Berghahn, 1–32.

Anderson, D.G. and Campbell, C. 2009. Picturing central Siberia: The digitization and analysis of early twentieth-century Central Siberian photographic collections. *Sibirica*, 8(2), 1–42.
Anonymous. 1941. Suglan (osedanie kolkhozov) [Assembly (sedentarization of collective farms)]. *Evenkiiskaia Novaia Zhizn'*, 23.
Argounova-Low, T. 2012. Narrating the road. *Landscape Research*, 37(2), 191–206.
Bloch, A. 2004. *Red Ties and Residential School: Indigenous Siberians in a Post-Soviet State*. Philadelphia, PA: University of Pennsylvania Press.
Bourdieu, P. 1977. *Outline of a Theory of Practice*. Cambridge: Cambridge University Press.
Brandišauskas, D. 2009. Leaving footprints in the taiga: Enacted and emplaced power and luck among Orochen-Evenki of the Zabaikal region in eastern Siberia. Ph.D. dissertation, University of Aberdeen. Publication forthcoming, Uppsala: University Press.
Campbell, C. 2003. Contrails of globalization and the view from the ground: An essay on isolation in east-central Siberia. *Polar Geography*, 27(2), 97–120.
Campbell, C., not dated. Mobilization and isolation as outcomes of a dysfunctional Soviet landscape. Paper presented at the Havighurst Center for Soviet and Post-Soviet Studies Annual Young Researchers Conference 'Russia in Global Context: Peoples, Environments, Policies,' Oxford, OH, 7–9 November 2003, Available at: http://www.units.muohio.edu/havighurstcenter/publications/documents/Campbell.pdf [accessed: 1 October 2012].
Dawson, P.C. 2003a. Analysing the effects of spatial configuration on human movement and social interaction in Canadian Arctic communities, in *Proceedings of the 4th International Space Syntax Symposium* [Online], 37.1–37.14. Available at: http://www.spacesyntax.net/symposia-archive/SSS4/fullpapers/37DawsonAnlypaper.pdf [accessed: 1 October 2012].
Dawson, P.C. 2003b. Examining the impact of Euro-Canadian architecture on Inuit families living in Arctic Canada, in *Proceedings of the 4th International Space Syntax Symposium* [Online], 21.1–21.16. Available at: http://www.spacesyntax.net/symposia-archive/SSS4/fullpapers/21Dawsonpaper1.pdf [accessed: 1 October 2012].
Donahoe, B. 2006. Who owns the taiga? Inclusive vs. exclusive senses of property among the Tozhu and Tofalar. *Sibirica*, 5(1), 87–116.
Duncan, J. 2010. *How Intelligence Happens*. New Haven, CT: Yale University Press.
Fondahl, G. 1998. *Gaining Ground? Evenkis, Land and Reform in Southeastern Siberia*. Boston: Allyn & Bacon.
Forsyth, J. 1992. *A History of the Peoples of Siberia: Russia's North Asian Colony, 1581–1990*. Cambridge: Cambridge University Press.
Habeck, J.O. 1998. *Seßhaftwerdung und Seßhaftmachung sibirischer Rentiernomaden. Siedlungsstruktur und Siedlungsgeschichte im Ewenkischen Autonomen Kreis* [Sedentarization of Siberian reindeer nomads: Settlement

structure and history in the Evenki Autonomous Okrug]. *Berichte aus dem Arbeitsgebiet Entwicklungsforschung am Institut für Geographie Münster*, 30, Münster: Westfälische Wilhelms-Universität Münster, Institut für Geographie.

Habeck, J.O. 2005. *What it Means to Be a Herdsman: The Practice and Image of Reindeer Husbandry among the Komi of Northern Russia*. Münster: LIT.

Habeck, J.O. 2006. Experience, movement and mobility: Komi reindeer herders' perception of the environment. *Nomadic Peoples*, 10(2), 123–41.

Hewes, G.W. 1955. World distribution of certain postural habits. *American Anthropologist*, 57(2,1), 231–44.

Hewes, G.W. 1957. The Anthropology of Posture. *Scientific American*, 196(2), 122–32.

Hewes, G.W. 1966. The domain posture. *Anthropological Linguistics*, 8(8), 106–12.

Ingold, T. 2011. *Being Alive: Essays on Movement, Knowledge and Description*. London: Routledge.

Istomin, K.V. and Dwyer, M.J., forthcoming. *Reindeer Herders' Thinking*. Oxford: Oxford University Press (under review).

Levin, M.G. and Potapov, L.P. (eds). 1956. *Narody Sibiri* [The Peoples of Siberia]. Moskva and Leningrad: Izdatel'stvo Akademii Nauk SSSR.

Levin, M.G. and Potapov, L.P. (eds). 1964. *The Peoples of Siberia*. Chicago: University of Chicago Press.

Mazzullo, N. 2005. *Perception, Tradition and Environment among Sámi People in Northeastern Finland*. Ph.D. dissertation, University of Manchester.

Povoroznyuk, O. 2007. Reindeer herders and hunters of eastern Siberia: Life of Kalar Evenks, in *International Handbook of Research on Indigenous Entrepreneurship*, edited by L.P. Dana and R.B. Anderson. Northampton: Edward Elgar, 137–54.

Safonova, T. and Sántha, I. 2011. Mapping Evenki land: The study of mobility patterns in eastern Siberia. *Folklore: Electronic Journal of Folklore* [Online], 49(3), 71–96. Available at: http://www.folklore.ee/folklore/vol49/evenki.pdf [accessed: 1 August 2012].

Sergejew [Sergeev], M. 1948. Die kleinen Völker des Nordens in der Epoche des Sozialismus [The small peoples of the North in the socialism era]. *Sowjetwissenschaft*, 1948(4), 29–74.

Simonova, V.V. 2007. TransSib: Put' v zhizni, zhizn' v puti [TransSib: Route in life, life en route]. *Sotsiologicheskie issledovaniia*, 2007(5), 103–13.

Sirina, A. 1995. *Katangskie Evenki v XX v.: Rasselenie, organizatsiia zhiznedeiatel'nosti* [Katanga Evenkis of the twentieth century and the ordering of their life-world]. Moskva: Institut etnologii i antropologii Rossiiskoi Akademii Nauk.

Sirina, A. 2006. *Katanga Evenkis of the Twentieth Century and the Ordering of Their Life-World*. Northern Hunter-Gatherer Research Series, 2, Edmonton: Canadian Circumpolar Institute Press.

Slezkine, Y. 1994. *Arctic Mirrors: Russia and the Small Peoples of the North*. Ithaca, NY: Cornell University Press.

Starrett, G. 1995. The hexis of interpretation: Islam and the body in the Egyptian popular school. *American Ethnologist*, 22(49), 953–69.

Tarasenkov, G.N. 1930. *Turukhanskii krai: Ekonomicheskii obzor s istoricheskim ocherkom* [Turukhansk Region: Economic overview with an historic sketch]. Krasnoiarsk: Izdatel'stvo Turukhanskogo RIKa.

Tugolukov, V.A. 1985. *Tungusy (evenki i eveny) srednei i zapadnoi Sibiri* [The Tungus (Evenkis and Evens) of central and western Siberia]. Moskva: Nauka.

Turov, M.G. 1990. *Khoziaistvo evenkov taezhnoi zony Srednei Sibiri v kontse XIX–nachale XX v.: Printsipy osvoeniia ugodii* [Evenki economy in the central Siberian taiga at the turn of the twentieth century: Principles of land use]. Irkutsk: Izdatel'stvo Irkutskogo gosudarstvennogo universiteta.

Turov, M.G. 2010. *Evenki Economy in the Central Siberian Taiga at the Turn of the Twentieth Century: Principles of Land Use*. Northern Hunter-Gatherer Research Series, 5, Edmonton: Canadian Circumpolar Institute Press.

Urry, J. 2007. *Mobilities*. Cambridge: Polity Press.

Uvachan, V.N. 1984. *Gody, ravnye vekam: Stroitel'stvo sotsializma na Sovetskom Severe* [Years like centuries: The construction of socialism in the Soviet North]. Moskva: Mysl'.

Vasilevich, G.M. 1969. *Evenki: Istoriko-etnograficheskie ocherki (XVIII–XIX v.)* [The Evenkis: Historic-ethnographical sketches (eighteenth–nineteenth century)]. Leningrad: Nauka, Leningradskoe otdelenie.

Vitebsky, P. 2005. *Reindeer People: Living with Animals and Spirits in Siberia*. London: Harper Collins.

Chapter 9
Shamanist Topography and Administrative Territories in Cisbaikalia, Southern Siberia

Joseph J. Long

Anthropologists have often explored the relationship between 'space' and 'place' in the way that social actors perceive and relate to the land. In scientific thought space is often characterized as something measurable, abstract, and even neutral, given meaning only by human activities and perceptual schemes (Casey 1996). In ethnographic accounts, state agencies are frequently described as treating land as neutral space in apportioning administrative territories. In these instances, the land is seen and managed as though from above through the technology of mapping (Bourdieu 1977, Gow 1995, Scott 1998). Places, on the other hand, are understood as meaningful phenomena in human experience, as 'nodes in the comings and goings of people, each characterised by its particular assemblage of relations' (Ingold 2000: 145). Phenomenological accounts of place focus less on territory, instead emphasizing 'landscapes' as place-filled environments that are experienced and constituted through moving and acting within them (Feld and Basso 1996, Hirsch and O'Hanlon 1995, Ingold 2000, Tilley 1994).

A distinction between territory as apportioned space and landscape as made up of meaningful places provides some useful insights into the ways in which state nationality policy and local ritual practices respectively have identified Buryat Mongols in Cisbaikalia with the land on which they dwell. Soviet policy-makers designated autonomous national territories to the indigenous peoples of Siberia and supported the development of conterminous, institutionalized cultures. Among local Buryats, however, I argue that an alternative sense of belonging to land has been effected through ritual practices relating to ancestral spirits and former pastoral migration cycles. I use the term 'shamanist topography' to describe the constellation of places in the landscape – pastoral routes, sacred mountains, offering sites, and ancestral villages – where these rites take place. This is contrasted with the material topography of the state that includes roads, schools, houses of culture, memorials, and statues. The term 'topography' can refer either to a collection of features in a certain locality, or to processes of representing them. In talking here of shamanist and state topographies I follow the first definition in order to denote complexes of places in the landscape. In discussing these phenomena I hope to demonstrate that both these constellations of places are inscribed as significant in ways that are not limited to description or graphic representation such as mapping.

In arguing that these two topographies together inform Buryats' experience of the Cisbaikal landscape, I therefore resist too simple a dichotomy between indigenous place and state space. Nevertheless, I illustrate that relationships to these topographies have historically been constituted in distinct ways. A recent shift by the Russian state from the principle of national territorial autonomy to a revived version of national cultural autonomy means, however, that both state agencies and local shamanists are now negotiating syncretic ways of identifying people with the landscape. This involves rendering the local shamanist topography visible to state agents and the wider public through conservation strategies, media coverage of ritual events, and the physical demarcation of sacred places. I ultimately argue that such creative strategies are necessary because neither the Soviet principle of territorial autonomy, nor a non-territorial form of cultural autonomy adequately grasp or account for indigenous Buryat understandings of belonging to the land.

Administrative Territories in Soviet and Post-Soviet Russia

State agents and institutions operating at a macro level often tend to see the world from above, perceiving land through the technology of mapping and treating it as apportionable space (Scott 1998). Working on the principle of centralized planning, the Soviet state did just this as it divided land into territorial packages. At the local level the state rationalized and controlled the economy of land use through state farms and herding collectives (Anderson 2000, Habeck, this volume, Humphrey 1998). At a federal level the same state managed regional government and the question of national self-determination by creating autonomous national territories (Hirsch 2005, Martin 2001, G. Smith 1996, J. Smith 1999).

Inheriting a vast, culturally diverse territory from the Russian Empire, the early Bolshevik government had to win the loyalty of national intelligentsias to govern effectively and relied on local elites to undertake infrastructural and cultural development at the regional level (Hirsch 2005, Martin 2001, G. Smith 1996, J. Smith 1999). The answer to 'the national question' within Russia was therefore to develop a federal system that incorporated national territorial autonomy. Indeed Stalin, as commissar for nationalities in the early Soviet government, believed that a national homeland was crucial to the definition of nationality. Stalin explicitly rejected Austrian Marxist Karl Renner's principle of 'national cultural autonomy' as a means of governance for multi-national states, instead favouring a principle of self-determination within national territories (Stalin 1936 [1913]). Renner, writing in 1899, had proposed a system in which state and nation are separated. In Renner's vision the state would be responsible for defence, law and economy, while national groups within the state would have control over their own educational and cultural affairs (Renner 2005 [1899]). Under Soviet federalism, however, national cultures were institutionalized to give identity to the political territories to which they were assigned. The Soviet federal system inherited by the Russian Federation in 1991 mixed purely administrative federal territories (*oblasts* and *krais*) with these national territories of varying size (autonomous republics, autonomous *oblasts* and autonomous *okrugs*). In practice the equation of people and territory was not perfectly neat. Some national minorities were left outside the homelands bearing their name and titular nationalities were not always a majority in their designated territories. The model of national territories was maintained, however, and a symbolic equation between people and territory was certainly made in Soviet policy, with members of titular national groups often disproportionately represented in local government (Martin 2001).

When it was founded in 1923 the Buryat-Mongolian Autonomous Soviet Socialist Republic (ASSR) incorporated much of Cisbaikalia, the area to the west of Lake Baikal, as well as areas of Transbaikalia to the east of the lake (Map 9.1). The development of a state infrastructure in the Republic was explicitly tied to a process of 'cultural-national development' and, in the 1920s and 1930s, Buryat literary, fine-art and performing art forms were developed in the institutions of the new territory (Badiev 1983). These new forms were propagated in schools, higher education colleges and houses of culture and were part of a broader drive for literacy, education and development under the socialist project of civilizing and modernizing indigenous peoples (see Grant 1995, Anderson 1996).

Map 9.1 Buryat-Mongolian ASSR 1923–1937

In the 1930s, however, Stalin, by then head of the federal Soviet government, undertook measures to undermine, repress and persecute national elites, fearful that developing national identification may threaten the hegemony of Soviet power and identity. Especially wary of pan-Mongolism in the Baikal region, Stalin divided the Buryat-Mongolian ASSR into three distinct national territories in 1937 (Balzer 1999). A shrunken Buryat republic remained, while in Cisbaikalia the Ust'-Orda Buryat-Mongolian National Okrug was established within the larger federal territory of Irkutsk Oblast (Map 9.2). Similarly, the former easternmost part of the Republic became the Aga Buryat-Mongolian National Okrug. In 1956 the *okrugs* lost the denomination *Mongolian*, and in 1978 were renamed as *autonomous* rather than *national* (Tarmakhanov, Damashek and Sanzhieva 2003).

Autonomous *okrugs* existed across Soviet Russia as the smallest form of national territories, all constituted within larger federal territories. Following the collapse of the USSR, however, the 1993 Russian constitution granted *okrugs* equal status with the other territories of the Russian Federation each with their own assembly (*duma*), their own representatives in the federal parliament, and autonomy over their budgets (Sakwa 2008). Under the principle of territorial autonomy, then, the Soviet state and its immediate successor in post-Soviet Russia followed the model of apportioning land as national space in the form of administrative territories. Below I undertake a shift of perspective to consider landscape-as-experienced to present a more nuanced understanding of Buryats' identification with the land than a simple equation between people and territory.

Ritual and Belonging in Cisbaikal Buryat Landscapes

Ethnographies of Northern Asia and the circumpolar North have made a considerable contribution to the anthropological literature on landscape in recent years, paying particular attention to the way that pastoralists, hunters and herders experience the environments through which they move (for example, Anderson 2000, Jordan 2011). These works have furthermore illustrated the ways in which incorporeal spirits are understood as active agents in the landscape, capable of influencing fortune and mediated through ritual practice. Caroline Humphrey (1995) has surveyed pastoral cycles and related offering rites among different Mongol groups, noting the way in which offering places that are met during the nomadic cycle in Mongolia become identified with the kin and clan groups that herd in these pastures. There, as in many contexts, ritual practices also confirm who belongs to family groups and herding collectives, cementing kin relations in the process. Although Cisbaikal Buryats largely live in a post-pastoral milieu – many now live in cities and large settlements away from their ancestral villages – pastoralism retains its significance through rituals that constitute a relationship with past dwelling places and migration routes. The shamanist topography – those sacred mountains, ancestral villages, roadside offering sites, and past migration routes – memorializes the pastoral history of kinship groups and still anchors a

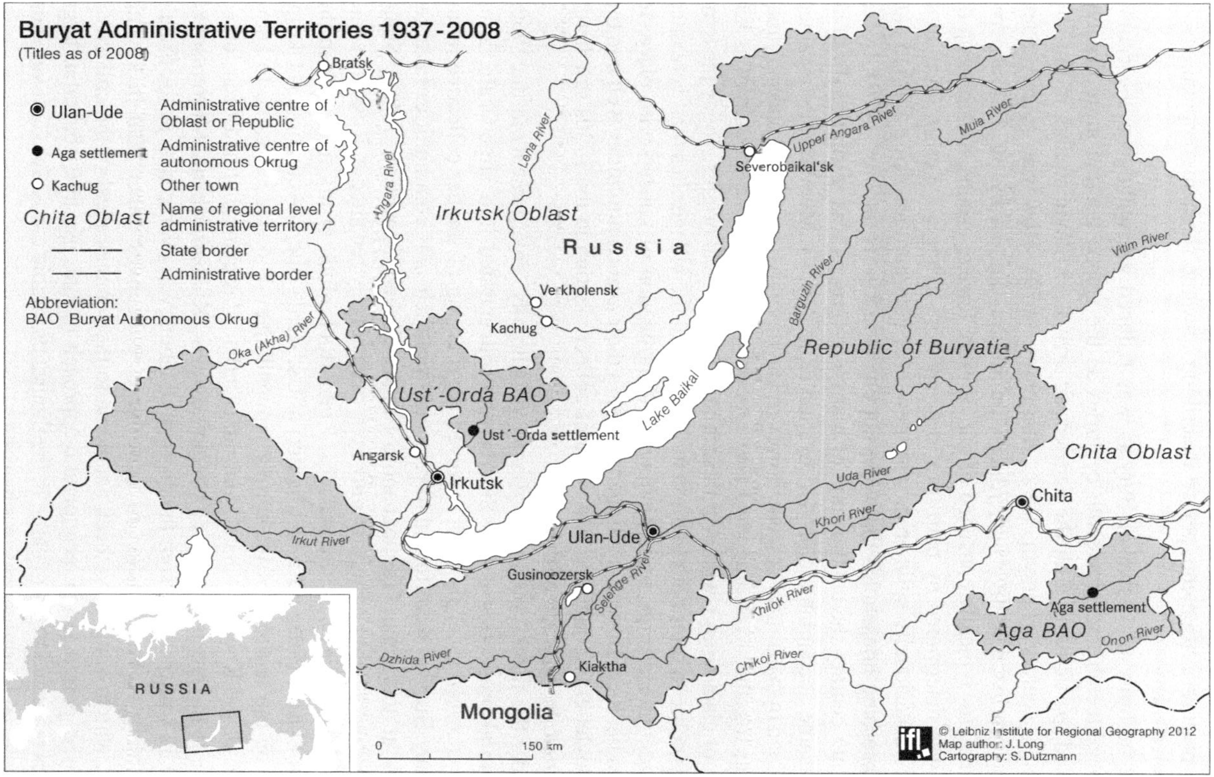

Map 9.2 Buryat administrative territories 1937–2008

sense of belonging to them. As I describe below, these rites define kinship relations and the obligations of reciprocity that they imply. It is therefore worth briefly sketching out the historical process through which Buryats moved from their former pastoral settlements before describing how the shamanist topography is constituted today.

The mobility of Cisbaikal Buryats was reduced gradually over the course of the eighteenth and nineteenth centuries as the steppe became more densely populated, particularly with the accelerated arrival of Slavic settlers in the late nineteenth century (Forsyth 1992, Humphrey 1979). However, at this point Buryats continued to live in herding communities known by the Buryat term *ulus*. The *ulus* typically consisted of one or more lineage groups, membership of which was determined by patrilineal descent from a single ancestor. Over time lineages occasionally splintered, with new *ulus* communities being formed and even absorbing households that were not cognatic kin (Humphrey 1979). In the nineteenth century the composition of many *ulus* groups was institutionalized as they became accountable units for taxation (Raeff 1956). Membership remained determined by patrilineal descent, however, and through herding and making ritual offerings together over time most *ulus* groups have come to be thought of as kin groups, or clans (*rody*), whatever their mixed genealogical origins. Today the term *ulus* has come to refer both to the clan itself and to the village sites that were once nomadic winter camps, underpinning an identification of kinship with place. In Cisbaikalia, local Buryats continued to live in their *ulus* communities well into the early Soviet era, though the first wave of collectivization in the 1930s saw herds designated the property of the state.

Under the federal policy of *ukrupnenie* (literally: enlargement) in the 1960s agricultural production across Siberia was concentrated into larger state farms known as *sovkhozy*. This often involved the merging of smaller collective farms and many Buryats were forced to relocate from their *ulus* communities to larger villages at this time (Tarmakhanov, Damashek and Sanzhieva 2003). Some *sovkhozy* were based on former *ulus* sites, occasionally renamed in Russian, some on the sites of Slavic settler villages. I undertook much of my fieldwork in the north of Ekhirit-Bulagat Raion (district) of Ust'-Orda Buryat Autonomous Okrug. Living on the territory of the former Novonikolaevsk *sovkhoz*, I came to know families of the Khertoi, Khara Nuur, and Darbai *ulus* clans that had been forced to move to the larger villages of Novonikolaevsk, Khabarovsk and Muromtsovka (Map 9.3). Under the *sovkhoz* the division of labour was overseen by state officials and herding was undertaken in these larger collectives rather than in kinship groups, which would have been anathema to the socialist ideal of the collective. The Soviet period saw a technocratic approach to land use in Cisbaikalia: large brigades managed herds of cattle in the thousands, arable crops were cultivated on the steppe, and farming was mechanized.

Since the last decades of the Soviet era, a steady migration out of the countryside has been underway as Buryats and Slavs alike have sought higher education and work in the cities of the region (Humphrey 1989). This process has

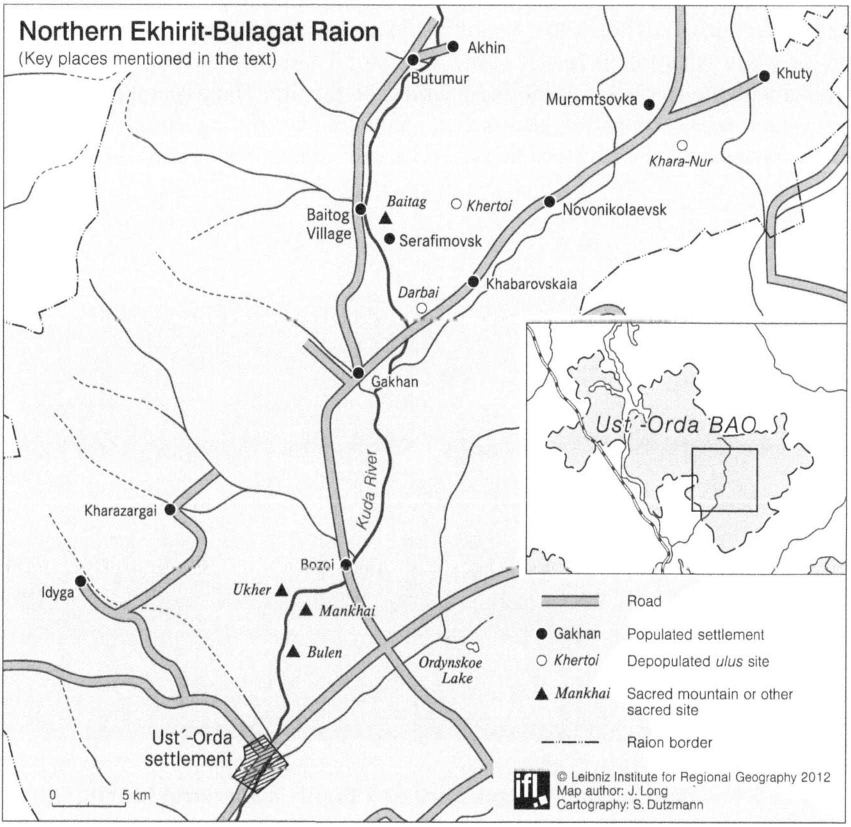

Map 9.3 Northern Ekhirit-Bulagat Raion

been accelerated since the collapse of collective farming in the 1990s and driven by the resultant rural unemployment. In the villages of the former Novonikolaevsk *sovkhoz* many houses now stand empty or are boarded up. The derelict tractor garages and cattle sheds on the edge of the village have gradually been plundered for building materials, and farm machinery lies rusting by the side of the roads. Yet despite the atomization of kin groups that has come about with processes of migration away from rural areas, belonging to *ulus* communities is maintained through the ritual constitution of kin relationships that still take place in what are often depopulated ancestral villages. Here only a brief sketch of these rites is possible, but I hope that it will be sufficient to illustrate the way in which they constitute a relationship to the landscape.

Each summer Buryats travel from the cities and villages of the region to their ancestral *ulus*. There they make offerings to the spirit of the family hearth, known as the *mongol-burkhan* in a rite known as 'feeding the *mongol*', and also to the ancestral spirits that continue to dwell there (see also Khandagurova 2008). These offerings are usually conducted by heads of households and prayers are made for the protection of all household members. In most cases this entails lighting a fire in the place where their family home once stood and sharing milk products and vodka with the spirits by sprinkling libations on the fire. The pastoral character of ancestors' relationships to place is also remembered: offerings are usually made both at the site of the ancestral hearth and at the summer pastures of forebears.

Figure 9.1 Feeding the Mongol spirit at a family's ancestral hearth, Khertoi *ulus*, June 2007

Source: Photograph by Joseph J. Long

Collective rituals are also undertaken annually by lineage groups and *ulus* clans. These rituals, known as *tailgan* rites, are important in constituting belonging to the clan (Humphrey 1979, Khandagurova 2008, Tugutov 1978). On these occasions kinsmen come together in their ancestral villages, some travelling many miles from the cities of Irkutsk, Ulan-Ude, or even returning from Moscow and Saint Petersburg. In these rites male clan members again offer milk and vodka libations at a ritual fire before slaughtering and butchering a sheep, sharing the cooked meat between corporeal clan members and also with incorporeal spirits on a specially constructed pyre. A designated clan elder, or a shaman, recites prayers for the offerings to be received and asks the spirits of the homeland – ancestors, deceased shamans and the master-spirit (*ezhen*) of the *ulus* – for the protection and good fortune of clan members. Kinship is firmly emplaced in the ancestral homeland, then, and a Buryat sense of belonging constitutes ties between kinsmen – both corporeal and incorporeal – and place.

Most ritual sites in Cisbaikalia are not marked in any way, they are simply known by local people, but in recent years there has been an increasing trend of erecting and dedicating ceremonial tethering posts, known as a *serge*, at the site of former family homes and clan offering sites. Among eastern Buryats of Transbaikalia and in Mongolia, a clan's designated offering place is usually marked by a stone cairn known as an *obo* (or *ovoo*). In Cisbaikalia, the *serge* is a more prevalent means of marking offering sites. *Serge* posts usually stand at about five feet tall with a pointed top and two or three horizontal grooves cut into them. Some Russian ethnographers have identified the *serge* as a phallic symbol (Skrynnikova et al. 2000), others have likened it to a 'tree of heaven' reaching from the earth to the sky (Zhambalova 2000). *Serge* posts that mark designated offering places are common to several of the Altaic speaking peoples of Siberia and in the Baikal region they were historically erected outside Buryat homes at the time of marriage to mark the presence of a new household (Zhambalova 2000: 262–74). Today *serge* posts are making a steady reappearance at the sites of ancestral villages and homes, a material and visible marker of belonging in the landscapes of Cisbaikalia.

In 2008 I joined the Kharazargai *ulus* for their summer *tailgan* in the Kuda river valley. After convening at the *ulus* winter settlement, still an occupied village, we

Figure 9.2 Dedicating a *serge* (ceremonial tethering post), Khuty, July 2006

Source: Photograph by Joseph J. Long

followed their ancestors' migration route to the summer pastures. The rite was held on Ukher, a hillock at the foot of the sacred mountain of Mankhai. The route was much slower than the asphalt road that could have been taken, but following the paths of the ancestors formed an important part of the rite.

Figure 9.3 A Kharazargai *ulus* elder leads a *tailgan* rite at Ukher, facing the sacred mountain of Mankhai, June 2008

Source: Photograph by Joseph J. Long

The journey from Kharazargai to the clan's *tailgan* site included stopping at a designated place along the road to offer libations of vodka. These offering places, known as *bar'sa* sites, occur at many places along the roads of Cisbaikalia. Offerings of tobacco or vodka are always made for safe passage when passing these sites on significant journeys such as travelling to clan rituals, weddings and even civic events such as folklore festivals or sports competitions. In everyday travel between villages, *bar'sa* (or *barisa*) sites are acknowledged by casting a little tobacco out of vehicle windows, sounding the horn of a vehicle or simply raising a hand in acknowledgement of the site. The historical reasons for the location of a particular *bar'sa* are occasionally lost or obscure to those simply passing a site. Some people saw these offerings as being for any spirits that may be encountered along the road. Other *bar'sa* sites are, however, dedicated to the master-spirits of sacred sites that are near to the road and passed on the journey, or to the spirits of deceased shamans whose funeral sites may be nearby. Certain springs, lakes and mountains in the landscape are home to particularly powerful spirits and treated with special reverence by all the Buryats of the area. Often the identity of the master-spirit and the place itself are merged, with ancestral names becoming toponyms. These sites, such as the mountains of Kapsal, Mankhai, Bulen and Baitag in the Kuda valley are the focal point of larger territorial groupings to which Buryats belong, home to common ancestors that unite *ulus* groups from across the area. While *ulus* and lineage *tailgan* rites were illegal but possible in the late-Soviet era, larger-scale *tailgan* rites, which were undertaken in the pre-Soviet era (see Khangalov 2004 [1958]), were impossible under Soviet atheism. Larger rites have been revived in recent years, however, as I discuss below.

As Buryats move through the Cisbaikal landscape they are reminded, at significant places, of their relationships with and obligations to the spirits that dwell there. This is experienced most strongly at annual *tailgan* rites. Central to all of this ritual practice is a principle of reciprocity – of making offerings in return for protection of the household and herd, good fortune for the clan, or for safe passage through the landscape.

Despite the atomization of kin groups in recent decades, the collapse of the Soviet economy has made kin networks more vital than ever for economic survival as rural and urban kinsmen provide one another with financial support, meat and dairy products, or labour in the peak hay-cutting season. These networks of support are constituted through the ritualized sharing of meat. Rituals of reciprocity and sharing inform a distinctly Buryat sense of belonging to people and place, an ethos not always accounted for in political conceptions of national belonging that make a more simplistic identification between people and land.

State and Shamanist Topographies in the Cisbaikal Landscape

In comparing political and indigenous ways of identifying people with land I do not wish to make too simple a dichotomy between Buryats as indigenous place-makers and the state as a manager of territory, for the state is also experienced at a phenomenological level. As well as collectivizing property and developing national cultures, one of the main tasks of regional government under Soviet rule was to develop the infrastructure of the state. This infrastructure not only involved roads, schools, factories and farms that were seen as the key to socialist development, but the institutions that gave territory both a Soviet, civic identity and a Buryat national identity: Lenin statues, war memorials, theatres, sports stadiums and houses of culture became the places in which belonging to civic communities was cemented and expressed. In what Francine Hirsch (2005) refers to as a 'double assimilation', Soviet institutions simultaneously propagated a sense of local national belonging and identification with the wider symbolism of Soviet culture. Folklore concerts in the local houses of culture, working parties that maintain village war memorials, and speeches made in front of government buildings on civic occasions have also given meaning to places in the Cisbaikal landscape, substantiating the state's own material topography of significant places.

In his ethnography of Evenki in Central Siberia, Nikolai Ssorin-Chaikov (2003) characterizes the taiga, into which Evenki go to hunt and herd reindeer, as a place beyond the material infrastructure of the state (albeit a place where the state has a reach through practices and discourses). This can be contrasted with the landscapes of the South Siberian steppe, however, where the shamanist topography is overlaid, or perhaps interwoven with the material topography of the state in Buryat experiences of landscape. *Bar'sa* sites appear on the asphalt roads that once brought supplies to collective farms, some *ulus* sites were developed as collective farm centres, and local Buryats move between and through these different kinds of places just as they might move between a civic role as a farm manager, teacher or local administrator to act as an officiating elder or shaman at a clan rite.

In a recent study of Zimbabwean understandings of belonging, Fontein (2011) suggests that acknowledging different and proximate material indices of belonging – in his example, tribal ancestral graves and the ruins of white colonial farming – does not necessarily mean that these complexes of signification exist on different ontological planes, or require a radical shift of perspective to perceive. Rather Fontein argues that they are often mutually constitutive in narratives of belonging, with ghosts and graves of white farmers incorporated into local mythologies. In Cisbaikalia civic and shamanist discourses of identification remained markedly distinct until recently. The Soviet persecution of religion meant that while civic place-making practices were public and open, shamanist ritual practices, though widely known about, were kept out of public view and were absent from ideologically acceptable discourses of identification. Although my interlocutors in the villages of Cisbaikalia remember even local officials undertaking offering rites, they were often carried out at night, or in shorter forms than the all-day *tailgan* rites that are possible today. Recently, however, Buryat shamanists have begun to collaborate with state agencies and others to create syncretic discourses and practices of place-making, to publicize and re-frame the shamanist topography as a focus for cultural heritage, conservation and tourism. This development has been necessitated and crystallized in part by the change in the Russian federal government's way of identifying people with land at a macro-level and a switch from an ideology of territorial autonomy to cultural autonomy.

The Shift to Non-Territorial Cultural Autonomy

Whatever the relationship of state and shamanist topographies at the local level, the principle of territorial autonomy at least provided a space in which traditional practices and relations to land could exist unhindered in the immediate post-Soviet era. The revival of interest in indigenous national culture in the 1990s, and the increasing presence of shamanist motifs and tropes in Buryat fine art, sculpture and performing arts gave image to the importance of shamanism and pastoralism to the fabric of life within that national space.

Yet this marriage of cultural revival and territorial sovereignty proved short-lived. From the mid-2000s onwards the Russian government under Vladimir Putin sought to simplify the Russian federal system by decreasing the number of subject territories in Russia (Oracheva 2009, Sakwa 2008). This process began with dissolving the autonomous *okrugs* in Siberia. In April 2006 a referendum was held on the merger of Ust'-Orda Buryat Autonomous Okrug with Irkutsk Oblast resulting in the unification of the two regions in January 2008. Rumours suggest that the process was less than democratic with voters intimidated and opposition silenced. Moreover, the pro-unification propaganda surrounding the referendum paid little attention to Buryat autonomy and focused instead on the economic regeneration for the whole region (see Graber and Long 2009). In the run-up to the referendum the administration of Irkutsk Oblast did, however, seek to reassure Buryats that there would still be institutional support for preserving Buryat culture even if the *okrug* was to be dissolved. The administration pointed to a 1996 federal law *On National Cultural Autonomy*[1] as the means through which Buryat culture would be safeguarded. When this law was passed by the Yeltsin government in 1996, it helped Russia to fulfil the Council of Europe's *Framework Convention for the Protection of National Minorities* (Bowring 2005: 191). Russia's 2000 report to the Council of Europe stresses the formation of a number of national cultural associations titled 'autonomies' under the law. Of the seven formed at the federal level by that time, five were nationalities without a given territory in Russia (Torode 2008). The formation of cultural autonomies in areas formerly politically autonomous is something of a new phenomenon, however. The key change locally is the fact that the principle of national cultural autonomy and the new funding arrangements for cultural activities have divorced them from the Soviet rubric of 'national self-determination' based upon territory. This rediscovery of Renner's principle has been the model through which culture has been decoupled from territory and moved into the institutions of civil society.

Bringing the Shamanist Topography into Public View

In the 1990s, following the collapse of the Soviet Union, Buryats across the Baikal region began to be more open about their shamanist beliefs and practices (Zhukovskaya 2000). In Ust'-Orda Buryat Autonomous Okrug large-scale *tailgan* rites began to be revived at this time in which Buryats from across the area come together to make offerings, often including the ritual slaughter of horses. The settings for these offering rites are usually those sacred mountains discussed above that are home to particularly powerful spirits.

In 1997 an association of shamans of Ust'-Orda Buryat Autonomous Okrug was founded and named *Sakhilgan* (Mikhakhanova-Baliueva 2006). The organization has conducted most of the larger rites in recent years and has increasingly engaged mass media to publicize its work as part of a concerted effort to make a Buryat understanding of landscape and ecological relations more visible in the public domain. Since the unification of *okrug* and *oblast* the association has declared its offering rites as being on behalf of 'all the residents of Cisbaikalia' (Kork 2008) and has often invited press and television journalists to their rituals. In rites that I

1 Russian Federal Law 76-F6 (1996), *O natsional'no-kul'turnoi avtonomii.*

attended in 2008 and 2011, several television companies and print journalists were present.

As a part of this process, articulation of an indigenous topography is increasingly being syncretized with discourses of cultural heritage and conservation along with a growing awareness of tourism. The dissolution of Buryat territorial autonomy has been a catalyst for this process, but a number of other developments have also influenced this discourse. These include illegal digging for stone on several sacred mountains, the proposed routing of a gas pipeline through the Kuda valley, and the planned development of a tourist fishing reserve in a sacred lake. I pay particular attention to the first example here as it is a problem that I have witnessed at close hand.

In summer 2007 illegal quarrying for stone was discovered on the mountain of Mankhai (Map 9.3). As well as being a sacred mountain and site of summer *tailgan* ceremonies for several local *ulus* clans, the mountain is home to ancient petroglyphs depicting horses. These carvings have been a focus of attention for several generations of Russian archaeologists and are thought to have been created between the eighth and fourteenth centuries (Nikolaev and Mel'nikova 2008). Several exemplars of the petroglyphs were removed by Soviet ethnographers in the 1950s, but of twenty-seven figures that remained all except six have now been destroyed by the digging. Some suspect that petroglyphs may even have been taken as trophy centrepieces for fireplaces in the newly built homes of the urban elite.

In September 2007 I was asked to participate in an expedition, funded by the *okrug* administration, to investigate the damage to Mankhai and to make recommendations for its preservation. The expedition presented an opportunity to gain media attention and inform people of the historical and spiritual significance of the mountain. Lobbying to mount the expedition had been orchestrated by Aleksei (pseudonym), a local shaman. Aside from Aleksei and myself, the delegation included two members of the *okrug* parliament, two scholars from the National Institute of the Humanities at Buryat State University in Ulan-Ude, archaeologists from Irkutsk State Technical University and the director of the Ust'-Orda Centre for Cultural Heritage. Before inspecting the damage Aleksei conducted an offering rite on the mountaintop for the success of the expedition and the successful preservation of the mountain. After lighting a ritual fire, he led prayers to the master-spirit of Mankhai and all present joined in with the familiar process of drinking and offering libations of milk and vodka for the spirits. Having visited the site two years previously it was a shock to see the damage to the mountainside: industrial-sized diggers had clearly been used to gauge out large trenches on a spot where I had previously seen a number of the petroglyphs.

When the expedition returned to Ust'-Orda settlement to draft a report to the *okrug* administration – which was still functioning at that time – members took it in turns to give their contributions to its findings. Aleksei explained the significance of Mankhai as a sacred place for his clan; the archaeologists present discussed the uniqueness and importance of the petroglyphs; the professors from Ulan-Ude emphasized that damage to Mankhai signified too little emphasis on Buryat

Figure 9.4 Surveying damage to Mankhai, September 2007

Source: Photograph by Joseph J. Long

national traditions and culture within *okrug* schools. Finally, the representative of the *oblast* tourist board was consulted about how a visitor centre might be promoted as part of the tourist route from Irkutsk to Baikal.

The resulting report proposed that Mankhai become a 'specially protected natural-cultural territory for the development of tourism' (Kompleksnaia Ekspeditsiia 2007). The purpose of the park was emphasized as 'the preservation of historical-cultural heritage of the native population of Irkutsk Oblast and Ust'-Orda Buryat Autonomous Okrug'. The proposal further recommended that the institution of a park would be 'the best method of protecting the natural and cultural complex and biological diversity', stressing a ban on building and digging on the territory and limiting use of resources in the park to 'organic' and 'traditional' methods. The position statement also emphasizes an onus on the state to provide funding for the park via the local *raion* administration. Accompanying the report was a map outlining the boundaries of the proposed park. Archaeologists, Aleksei and representatives of the administration worked together to ensure that the territory included the sites of all remaining petroglyphs, the ritual site of Ukher at the foot of Mankhai and the pastureland along the Kuda river valley.

Traditional land use and nature conservation have often been historically related concepts in Russian thinking. The Baikal region has a long history of creating state reserves (*zapovedniki*) dating back to 1916 (Metzo 2003: 69–70).

The first was at Barguzin in Transbaikalia, founded to preserve stocks of fur-bearing animals and so limiting hunting to indigenous methods. Increased nature and heritage conservation has, moreover, gained momentum in the region in recent decades. The Tunka Valley, a district of Buryatia to the south of Irkutsk, became a national park in 1991 (Metzo 2003), and the listing of Lake Baikal as a UNESCO world heritage site in 1996 has also increased awareness of global environmental and conservationist discourses in the region.

I noted above that state policies tend to rationalize land as *space*, apportioning territories for economic use, or as national homelands. If national territories no longer provided a symbolic space in which local relationships with ancestral homelands could be protected, it seems that to protect Mankhai as a place, Aleksei and his colleagues needed to find another way of placing the mountain under the protection of the state. A national park appeared to be an appropriate form of territory in which to contain Mankhai once the larger protective shell of national territory had disappeared. Moreover, in proposing the establishment of a national park, expedition members collaborated in creating a map of the proposed territory, marking on it the sites of petroglyphs and sacred places. Engaging with state agents involved re-presenting elements of the shamanist topography through a medium of state governance – defining Mankhai as a protected territory meant engaging statist technologies for imagining space.

This is not to say that the national territories had always been able to wield complete autonomy, nor that the Ust'-Orda Buryat Autonomous Oblast government had always proved successful at preserving Buryat cultural heritage. Incremental damage to Mankhai had been occurring for some years and Buryat language education decreased in the late Soviet era. I do, however, think that the symbolic loss of autonomy galvanized activists. Urgently gaining a resolution from the *okrug* administration before it was dissolved in January 2008 would at least allow local actors to subsequently lobby the *oblast* government in Irkutsk. It also gave local activists a form in which to assert a relationship with the land after opposition to unification had been so effectively muted during the referendum campaign.

What is particularly striking in this instance is the role Aleksei played. Whilst drawing together conservationist, archaeological and touristic discourses, Aleksei also underlined the position of local Buryats, shamans in particular, as mediators with master-spirits of place, drawing representatives of academic and state institutions into rituals of reciprocity with the spirits of Mankhai. Concurrently with his work on Mankhai, Aleksei acted as a local representative to an oil company that was planning a pipeline through the Kuda valley to join the city of Angarsk to pipelines to the north of Lake Baikal. Again the concern among local people was that the route would disturb sacred places, provoking anger among the spirits. Aleksei worked to map out sacred places that the pipeline route should avoid, again translating the shamanist topography into a form favoured by state agents and mineral companies. The pipeline's route was later moved altogether due to more public environmentalist concerns about its proximity to Baikal and the danger of leaks polluting the lake.

The *Sakhilgan* association has also played an increasing role in publicizing the sacred nature of certain local sites. In 2011 a number of shamans led offering rites at Ordynskoe Lake, another sacred location, to mobilize opposition to the development of a tourist fishing complex there. In the same year attempts to raise awareness of the local shamanist topography saw the installation and ritual dedication of shelters, picnic tables and *serge* posts at three important *bar'sa* sites along the tourist route from Irkutsk to the eastern shore of Baikal. Signs erected at these sites ask passing travellers to show their respect for local tradition and to make offerings for the master-spirits of each place.

All these strategies not only bring the material topography of shamanist places into view, but emphasize the principle of reciprocity as defining a relationship with the landscape. In publicizing local offering rites, including state officials in

Figure 9.5 Shamans from the *Sakhilgan* association lead offerings at a *tailgan* rite, June 2011

Source: Photograph by Joseph J. Long

ritualized commensality, and encouraging offerings from tourists passing sacred places, Buryat activists emphasize a local ethos of reciprocity with the spirits of the landscape. An extractive, exploitative approach to the landscape characterized by digging for stone, mineral extraction or industrial-scale fishing not only contradicts this indigenous relationship to landscape, but threatens those material places where local Buryats carry out relations with the spirits. Engaging the media to showcase these ritual practices also places Buryats, and particularly Buryat shamans, in the position of mediators with the powerful spirits that inhabit the Cisbaikal landscapes.

Concluding Remarks

The Soviet identification between people and territory was mediated through an official, institutionalized form of national culture, housed in the material infrastructure of the state – theatres, schools, and houses of culture. At the same time management of the land was technocratic and organized along civic lines in contrast to Buryat kin-based property and herding practices. Moreover, official Soviet culture was purged of the shamanist practices that constitute kinship, and a relationship with the landscape that is mediated through reciprocity with the spirits that dwell there.

Recent moves to bring the shamanist topography that has long existed outside of state-sanctioned practices into public view reflects an increasing willingness by shamanists to engage with state agents. Discourses of conservation are invoked to protect the integrity of this topography, and activists such as Aleksei undertake mapping exercises to record it in a graphic representational form. These practices and publicity strategies show us that neither a simplistic identification between people and territory, nor the currently favoured principle of national cultural autonomy, adequately acknowledge Buryat understandings of a relationship to the landscape as reciprocal, rooted in a shamanist topography that memorializes the pastoral history of their ancestors. Renner's original formulation of national cultural autonomy envisages a legislative division between state and nation: the state takes care of 'economic policy and material life' while the nation takes care of 'intellectual and cultural life' (Patton 2005: 114). This dichotomy between materiality and culture has been critiqued by Patton (2005) as an anathema to the worldviews of many indigenous peoples in which material places are integral to cultural practices. With the decoupling of culture from territory under the Russian policy of national cultural autonomy, Buryats are finding alternative discourses and practices through which to publicly assert attachment to places and to negotiate for different kinds of territories in which relations with the landscape can flourish unhindered. An ethnographic approach to the study of landscape that accounts for the ways in which local people constitute and identify with places allows us to understand how this process takes place and why it is held to be of such vital importance.

Acknowledgements

The doctoral fieldwork upon which this chapter draws was generously funded by the Social Science and Humanities Research Council of Canada's Baikal Archaeology Project. I am also grateful for the practical and administrative support provided by Artur Kharinskii and his colleagues at Irkutsk State Technical University during my stay in Cisbaikalia. I would like to thank Kirill Istomin and Jenny Peachey as well as two anonymous reviewers for helpful comments on the first draft of this chapter.

References

Anderson, D.G. 1996. Bringing civil society to an uncivilised place: Citizenship Regimes in Russia's Arctic Frontier, in *Civil Society: Challenging Western Models*, edited by C.M. Hann and E. Dunn. London: Routledge, 99–120.

Anderson, D.G. 2000. *Identity and Ecology in Arctic Siberia: The Number One Reindeer Brigade*. Oxford: Oxford University Press.

Badiev, A.A. 1983. *Kul'turnoe stroitel'stvo v Buriatskoi ASSR: Dokumenty i materialy* [Cultural construction in the Buryat Autonomous Soviet Socialist Republic: Documents and materials]. Ulan-Ude: Buriatskoe knizhnoe izdatel'stvo.

Balzer, M. 1999. Dilemmas of Federalism in Siberia, in *Center-Periphery Conflict in Post-Soviet Russia*, edited by M.A. Alekseev. Basingstoke: Macmillan, 131–66.

Bourdieu, P. 1977. *Outline of a Theory of Practice*. Cambridge: Cambridge University Press.

Bowring, B. 2005. Burial and resurrection: Karl Renner's controversial influence on the 'national question' in Russia, in *National Cultural Autonomy and Its Contemporary Critics*, edited by E. Nimni. London: Routledge, 191–203.

Casey, E. 1996. How to get from space to place in a fairly short stretch of time: Phenomenological prolegomena, in *Senses of Place*, edited by S. Feld and K. Basso. Santa Fe, NM: School of American Research Press, 13–52.

Feld, S. and Basso, K.H. (eds). 1996. *Senses of Place*. Santa Fe, NM: School of American Research Press.

Fontein, J. 2011. Graves, ruins and belonging: Towards an anthropology of proximity. *Journal of the Royal Anthropological Institute*, 17, 706–27.

Forsyth, J. 1992. *A History of the Peoples of Siberia: Russia's North Asian Colony, 1581–1990*. Cambridge: Cambridge University Press.

Gow, P. 1995. Land, People, and Paper in Western Amazonia, in *The Anthropology of Landscape: Perspectives on Place and Space*, edited by E. Hirsch and M. O'Hanlon. Oxford: Clarendon, 43–62.

Graber, K. and Long, J. 2009. The dissolution of the Buryat autonomous okrugs in Siberia: Notes from the field. *Inner Asia*, 11(1), 147–55.

Grant, B. 1995. *In the Soviet House of Culture: A Century of Perestroikas*. Princeton, NJ: Princeton University Press.

Hirsch, E. and O'Hanlon, M. (eds). 1995. *The Anthropology of Landscape: Perspectives on Place and Space*. Oxford: Oxford University Press.

Hirsch, F. 2005. *Empire of Nations: Ethnographic Knowledge and the Making of the Soviet Union*. Ithaca, NY: Cornell University Press.

Humphrey, C. 1979. The uses of genealogy: A historical study of the nomadic and sedentarised Buryat, in *L'Ecologie et Anthropologie des Sociétés Pastorales/ Pastoral Production and Society*, edited by l'Equipe écologie et anthropologie des sociétés pastorales. Cambridge: Cambridge University Press, 235–260.

Humphrey, C. 1989. Population trends, ethnicity and religion among the Buryats, in *The Development of Siberia: People and Resources*, edited by A. Wood and R.A. French. Basingstoke: Macmillan, 147–76.

Humphrey, C. 1995. Chiefly and shamanist landscapes in Mongolia, in *The Anthropology of Landscape: Perspectives on Place and Space*, edited by E. Hirsch and M. O'Hanlon. Oxford: Clarendon Press, 135–62.

Humphrey, C. 1998. *Marx Went Away – But Karl Stayed Behind: Updated Edition of Karl Marx Collective: Economy, Society and Religion in a Siberian Collective Farm*. 2nd Edition. Ann Arbor: University of Michigan Press.

Ingold, T. 2000. *The Perception of the Environment: Essays in Livelihood, Dwelling and Skill*. London: Routledge.

Jordan, P. 2011. *Landscape and Culture in Northern Eurasia*. Walnut Creek, CA: Left Coast Press.

Khandagurova, M.V. 2008. *Obriadnost' kudinskikh i verkholenskikh buriat vo 2 polovine XX veka (basseinov verkhnego i srednego techeniia rek: Kuda, Murino i Kamenka)* [Rites of the Buryats of Kuda and the Upper Lena in the second half of the twentieth century (in the catchment areas of the upper and middle reaches of Kuda, Murino and Kamenka)]. Irkutsk: Amtera.

Khangalov, M.N. 2004 [1958]. *Sobranie sochinenii*, vol. 1. Ulan-Ude: Buriatskoe knizhnoe izdatel'stvo.

Kompleksnaia Ekspeditsiia v Mestnost' 'Mankhai' 2007. *Polozhenie: Ob osobo okhraniaemoi prirodno-kul'turnoi territorii dlia razvitiia turizma* [Status report: On the specially protected natural-cultural territory for the development of tourism] [expedition report presented to the parliament of the Ust'-Orda Buryat Autonomous Okrug, 14 September 2007].

Kork, B. 2008. Rodovoi obriad vozle sviashchennoi gory. *SM Nomer Odin* [Irkutsk-based journal], 24 (1526), 19 June 2008.

Martin, T. 2001. *The Affirmative Action Empire: Nations and Nationalism in the Soviet Union, 1923–1939*. Ithaca, NY: Cornell University Press.

Metzo, K. 2003. *It Didn't Use to Be this Way: Household, Resources and Economic Transformation in Tunka Valley, Buriatia, Russian Federation*. Ph.D. dissertation, Indiana University.

Mikhakhanova-Baliueva, G.A. 2006. *Ekhered-Bulgadai Baabai* [The Forefather of the Ekhirit-Bulagat]. Ust'-Orda: Reprotsentr A1.

Nikolaev, V. and Mel'nikova, L.V. 2008. *Petroglify kudinskoi doliny* [Petroglyphs of the Kuda valley]. Irkutsk: Izdatel'stvo Irkutskogo gosudarstvennogo tekhnicheskogo universiteta.

Oracheva, O. 2009. Unification as a political project: The case of Permskii Krai, in *Federalism and Local Politics in Russia*, edited by C. Ross and A. Campbell. London: Routledge, 82–105.

Patton, P. 2005. National autonomy and indigenous sovereignty, in *National Cultural Autonomy and Its Contemporary Critics*, edited by E. Nimni. London: Routledge, 112–23.

Raeff, M. 1956. *Siberia and the Reforms of 1822*. Seattle: University of Washington Press.

Renner, K. 2005. State and nation (1899), in *National Cultural Autonomy and Its Contemporary Critics*, edited by E. Nimni. London: Routledge, 16–47.

Sakwa, R. 2008. *Russian Politics and Society*. 4th Edition. London: Routledge.

Scott, J.C. 1998. *Seeing Like a State: How Certain Schemes to Improve the Human Condition Have Failed*. New Haven, CT: Yale University Press.

Skrynnikova, T.D., Batoeva, D.B., Galdanova, G.R. and Nikolaeva, D.A. 2002. *Obriady v traditsionnoi kul'ture Buriat* [Customs in the traditional culture of the Buryats]. Moskva: Vostochnaia Literatura.

Smith, G. 1996. The Soviet state and nationalities policy, in *The Nationalities Question in the post-Soviet States*, edited by G. Smith. London: Longman, 2–22.

Smith, J. 1999. *The Bolsheviks and the National Question, 1917–23*. Basingstoke: Macmillan.

Ssorin-Chaikov, N. 2003. *The Social Life of the State in Subarctic Siberia*. Stanford: Stanford University Press.

Stalin, J. 1936 [1913]. Marxism and the national question, in *Marxism and the National and Colonial Question: A Collection of Articles and Speeches*. London: Lawrence and Wishart, 3–61.

Tarmakhanov, E.E., Damashek, L.M. and Sanzhieva, T.E. 2003. *Istoria Ust'-Ordynskogo Buriatskogo Avtonomogo Okruga: Uchebnoe posobie dlia obshcheobrazovatel'nykh uchebnykh zavedenii* [History of the Ust'-Orda Buryat Autonomous Okrug: Textbook for primary schools]. Ulan-Ude: Buriatskii gosudarstvennyi universitet.

Tilley, C.Y. 1994. *A Phenomenology of Landscape: Places, Paths, and Monuments*. Oxford: Berg.

Torode, N. 2008. National Cultural Autonomy in the Russian Federation: Implementation and Impact. *International Journal on Minority and Group Rights*, 15(2–3): 179–93.

Tugutov, I. 1978. The tailagan as the principle shamanistic ritual of the Buriats, in *Shamanism in Siberia*, edited by V. Diószegi and M. Hoppál. Budapest: Akadémiai Kiadó, 267–79.

Zhambalova, S.G. 2000. *Profannyi i sakral'nyi miry ol'khonskikh buriat (XIX–XX vv.)* [The secular and the sacred world of the Olkhon Buryats (ninetheenth–twentieth century)]. Novosibirsk: Nauka.

Zhukovskaya, N.L. 2000. Neo-shamanism in the context of the contemporary ethno-cultural situation in the Republic of Buryatia. *Inner Asia*, 2(1), 25–36.

[illegible] Nationality, regionalism and Islam [illegible]

[illegible]

[illegible] 2000. [illegible] Central Asia. [illegible] 1997.

Chapter 10

From Invisible Float to the Eye for a Snowstorm: The Introduction of GPS by Nenets Reindeer Herders of Western Siberia and Its Impact on Their Spatial Cognition and Navigation Methods

Kirill V. Istomin

The problem of the relation between technological and cultural change is probably one of the most discussed in the history of 'Western' thought since the renaissance epoch. It can indeed be argued that the special interest in this issue is a logical and important distinguishing characteristic of the European technological civilization. Although this problem is addressed most directly by philosophers and historians, its different aspects feature highly in the research agendas of almost all social and even some natural sciences. Recently, a whole interdisciplinary field of research, usually referred to as Science and Technology Studies (STS), has emerged to address this and related matters (see, for example, Bauchspies, Croissant and Restivo 2005).

In anthropology, the problem's aspect that has perhaps been most debated theoretically and researched empirically, is the question of technological innovations: how and under which conditions these innovations are created and/or adopted by people of particular cultures, how the adoption or disregard of the innovations influences other aspects of their culture and/or society, and which factors determine the outcome of these processes. It would be impossible to give even a minimally complete review of this debate here. It is important to mention, however, that the positions most often taken in this debate until the middle of the twentieth century can be roughly divided into two groups. The first group can be subsumed under the title of techno-economic causality or technological determinism and can be traced back to the Marxist tradition (e.g., Engels 1972 [1884]). Researchers from this group generally thought that technological innovations have pronounced formative effects on other aspects of culture (including social organization) and produce profound changes therein. The opposite influence, i.e., the role culture and society play in determining which innovations prevail and how they are adopted, was played down or was sometimes even denied. It seems most researchers from this group believed, implicitly or, sometimes, quite explicitly, that any technological innovation which is useful in a given ecological, infrastructural and historical context, is obligatorily adopted by people at least in the long term; although the social and/or cultural transformations, which can be related to this adoption, may affect the speed of the innovation's spread within society. Among anthropologists, positions of this kind have been most consistently taken by cultural materialists, such as Leslie White (1949), Marvin Harris (2001 [1968]) and Marshal Sahlins (2000).

The second group of theoretical positions on the issue can be subsumed under the title of sociocultural or 'mentalistic causality' (Pelto 1987 [1972], see below). The researchers holding these positions generally emphasized the causal role of cultural and societal factors in technological change. The extreme position of this kind would be that a technological innovation, however useful it can be from the point of view of economy and/or ecology, is not endorsed in a society unless it fits the cultural and/or social configuration already present therein. Cultural and social factors, therefore, greatly determine which technological innovations are adopted, how they spread in the society, how they are put to use, and what effect they have in terms of the economy, politics, and the way of life. The transformative effect of technological innovations, once endorsed, upon culture and society was not denied by most of researchers from this group. However, this effect was rarely focused on. The positions of sociocultural causality towards the technological change were most consistently taken by so-called cultural idealists, most notably Alfred Kroeber (1952) and Gregory Bateson (1967 [1935], 2000 [1972]).

The significant shortcomings of both groups of theoretical positions described were recognized by many researchers, including many of those who actually held them. These shortcomings became especially visible from empirical studies on adoption and effects of technological innovations, which proliferated in the second half of the last century. These studies provided numerous examples of both techno-economic causality of social and cultural changes and sociocultural causality of technological changes (or the lack thereof). It has been strongly felt, therefore, that the truth lay somewhere between the theoretical positions described. For example, Pertti Pelto, in his classic research on the adoption and effects of technological innovations among Sámi reindeer herders, concluded that:

> Neither 'techno-economic determinism' nor 'cultural causation' provides an adequate model for the complex feedback of effects in the human adaptational system. To my way of thinking, the juxtaposition of techno-environmental and 'mentalistic' causation is precisely where the logic of an ecological approach must lead us. The complex of interactive forces in the behavioral field includes all these extra-personal elements – physical and social environment, technological inputs, social organization, idea systems, and the bio-psychological characteristics of the individuals. (Pelto 1987: 179)

However, following this insight and constructing a workable theoretical model to explain adoption and the effects of technological innovations in such a way as to incorporate both techno-economic and sociocultural causations has proven to be a very difficult task. Indeed, such a model should necessarily specify conditions under which each kind of causation either becomes or ceases to be operative: something that would clarify the empirically observable differences in the process of adopting innovations as well as its results across cultures and time periods. Seemingly Pelto himself tended to think that these conditions were connected with the nature of innovations themselves. Thus he differentiated between innovations that directly affect the production process and, therefore, change the pattern of energy flows inside the community as a socio-ecological and socioeconomic system, and those which, despite changing the everyday life of people, do not interfere with the production process and the sociocultural adaptation of the community to its environment in general. Snowmobiles are an example of the former kind of innovation: in the case of reindeer herding communities, snowmobiles significantly alter the workload involved in herding operations as well as the relations between herders and reindeer and create dependency on the part of the herders upon outside sources of energy (gasoline). Kerosene lamps, radios and TV sets are examples of the latter. In accordance to Pelto, techno-economic causation works in the case of the first type of innovation. Indeed, their penetration into a society seriously affects the relative chances for the production success of separate productive units (e.g., families, households, etc.) creating what Pelto names 'techno-economic differentiation' (Pelto 1987: 168–9). Therefore, the productive units have to adopt the innovation whatever their culturally or socially induced attitudes towards it may be, simply in order to compete with other productive units. This causes more or less the obligatory adoption and spread of the innovation inside the society Furthermore, this kind of innovation is likely to have a huge transformative impact upon cultural assumptions and social relations inside the society through the direct effect on relative positions of productive units and economic relations between them as well as between the society and the outside world ('de-localisation') (Pelto 1987: 165–78). The innovations in the second group, on the other hand, are not likely to create significant techno-economic differentiation and their adoption, spread and social and cultural effects depend on the sociocultural conditions already present in the society. Hence, sociocultural causation is to be expected in this case (Pelto 1987: 166–7).

The other logical way to specify the conditions determining the relative importance of the techno-economic and sociocultural causations is to look at the cultures adopting the innovations rather than at the innovations themselves. This way of thinking has recently been explored by Thomas Hughes (2006), who differentiated between nature-based and technology-based cultures. Nature-based cultures are those without complicated technologies, depending mostly on processes naturally occurring in the environment ('natural forces' as Hughes names them) for their existence. These are chiefly cultures of hunters, gatherers, pastoralists and pre-industrial agriculturalists, who are greatly affected by the changes in natural forces. In technology-based cultures, on the other hand, complicated technology mediates between people and their environment altering and modulating natural processes. In these cultures, which include all the industrial and post-industrial cultures, the well-being of the people depends more on the smooth functioning of technology than on the changes in natural forces. According to Hughes, techno-economic causation is characteristic for technology-based cultures – explicable by the very special role technology plays in supporting their stability. In the case of nature-based cultures, sociocultural causation should be expected.

Unfortunately, these models also have significant shortcomings. First of all, the distinction between the two types of technological innovations proposed by Pelto is by no means neat and clear, and the same can be said about the distinction between nature- and technology-based cultures made by Hughes. Indeed, there is a big class of technological innovations which do not have a pronounced direct impact on the production process, but still have a huge indirect impact if used in a certain way (the telephone acts as a good example here). Similarly, the role of technology in the stable functioning of different cultures can be rather different and it is difficult to point out exactly when a nature-based culture becomes technology-based. Furthermore, both models still leave a lot of cases unexplained. Thus, the model of Hughes, based on cultural types, cannot explain the techno-economic causation certainly involved in the snowmobile revolution described by Pelto for Sámi, whose culture before this revolution was nature-based by all meaningful standards. On the other hand, the model of Pelto cannot explain the numerous examples of very slow and clearly culturally conditioned spread of many important technological innovations in the production sphere observed in the history of mankind. The history of European agriculture can serve as a good example here (Bloch 1966). This illustrates that extensive work still needs to be done on the theoretical model explaining the adoption and effects of innovations.

In my opinion, the best way to proceed with this work is through a careful and detailed study of the processes of adoption and spread of innovations in a particular culture. Only a study of this kind can demonstrate which factors are actually important in the causation of the technological innovation's spread and how they impact the process. It is exactly a study of this kind that is reported in this chapter. The aim of this study is to describe and analyse the introduction of one particular technology – the global positioning system (GPS) – by reindeer-herding Nenets of northwestern Siberia, whose life heavily depends on travelling skills. Admittedly, GPS is a product of industrialized society and a study of its spread among the reindeer-herding nomads can tell little about how technological innovations are created. However, such a study can reveal a lot about how technological innovations are adopted and spread within a society and, therefore, contribute to the theoretical debates described above. Indeed, GPS is a technology which clearly falls into the grey zone between the two kinds of technological innovations described by Pelto. This technology is not directly involved in the production process but, taking the studied group's way of life into account, it is clearly related to the production process and can affect it in numerous ways. Besides, the reindeer-herding nomads are by now included in industrial society and, therefore, their culture falls into the grey zone between the nature-based and technology-based cultures described by

Hughes. Therefore, this study shows both how the factors already described and theoretically assessed act in the so-called grey cases (in the final instance, the majority of the cases of adoption of technological innovations are likely to fall in this category) and, more importantly, what other factors, not yet theoretically assessed, play a role in such cases.

Another reason that makes the case of GPS introduction among Nenets particularly interesting is that this introduction could be observed and documented from its very beginning. Indeed, although the use of GPS for civil purposes was allowed as early as in 1983, the spread of GPS technology in Russia has effectively been hindered by Russian legislation, inherited from Soviet times and including restrictions on public access to any kind of satellite-produced data as well as on the use of high accuracy navigation equipment. Although these restrictions were admittedly the result of Soviet spy paranoia, they remained in force for almost 15 years after the collapse of the Soviet Union and were only completely lifted on 1 December 2006. Only after that the story of GPS adoption by indigenous groups living in Russia could really begin. As my field research among the Nenets covered the period from 2005 till 2011, this story could be observed in its entirety.

The main theoretically significant result, which I hope to demonstrate by this study, is that the adoption and spread of technological innovation is not a coherent process. It includes several phases, and the factors at work during each of them as well as the type of causation involved are likely to differ. The innovation's social and cultural impact essentially depends on how these phases are passed through and how the transition from one phase to the other occurs.

Region and Context of the Study

The region in which this study was conducted is the so-called Taz Tundra situated in the southern part of the Gydan Peninsula, northwestern Siberia. This region administratively represents the southern part of the Tazovskii Raion of the Yamal-Nenets Autonomous Okrug, Russian Federation.

The indigenous population of the region is known in anthropological literature as Nenets or Samoyeds. Economically, this population is divided into two groups: nomadic reindeer herders (*ty perta'* in Nenets) and semi-nomadic fishermen *(hala iorta'* in Nenets). The boundaries between the groups are by no means rigid and it is usual for separate households to switch from one group to the other. Nomadic reindeer herders live in camps, which usually consist of four to seven families (i.e., households). Each nomadic camp migrates throughout the whole year inside its own piece of territory (Dwyer and Istomin 2008, Istomin and Dwyer 2010). Although most of the reindeer-herding households currently own snowmobiles, draft reindeer and reindeer sledges are used during migrations as well as for some herding operations. In contrast to the reindeer herders, the semi-nomadic fishermen do not form camps. During most of the year, from September until the end of June or the beginning of July, each fishing household lives separately on its fishing

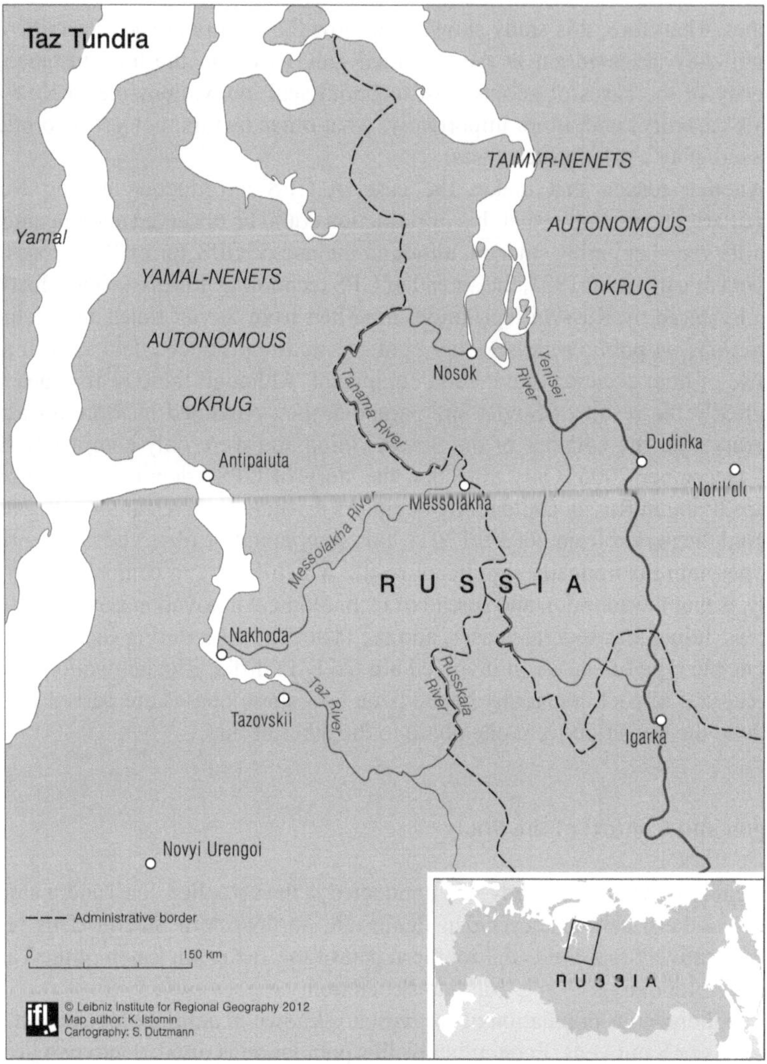

Map 10.1 Taz Tundra

grounds (*mara*) inside the Taz river flood plain and practices both open water and ice fishing with fishing nets. From July to August most of the households migrate by boats down the Taz river to the coast of the Kara sea to participate in the communal summer sea fishing (Donahoe and Istomin 2007).

In contrast to the nomadic reindeer herders, whose economy is oriented towards internal consumption, the semi-nomadic fishermen are mainly oriented towards the market. By far the largest part of their yearly catch (including almost the whole summer catch) is bartered with Russian settlers in exchange for goods and staple food. The fishermen usually master the Russian language much better than the herders and use much more technological innovations in their everyday life. Thus, most of them not only own snowmobiles, but also motor boats; and both means of transportation are used extensively depending on the period of the year. Many fishing households also have small herds of reindeer, which they entrust to nomadic reindeer herders for pasturing against payment. In the period from January to April, when fishing is not productive due to the thick ice cover, many such families reclaim their reindeer from the herders and migrate with them to the proximity of their fishing grounds.

Currently, there are three permanent settlements in the region: the town of Tazovskii, and the villages of Gaz-Sale and Nakhodka. The population of these settlements consists predominantly of Russian-speaking newcomers. The town of Tazovskii is especially important as a trading post between the nomadic and semi-nomadic population of the tundra and the sedentary world.

The data for this study was collected by the author within the framework of two successive research projects.[1] While the first research project focused on spatial perception and traditional methods of spatial orientation of tundra dwellers, the second project specifically concentrated on the effects of new technologies, with GPS devices playing the main role, on the spatial cognition of the local population. The research projects included seven fieldwork periods, each of them lasting two and a half to three months with intervals of approximately three months. The basic methods and techniques employed in the course of data collection included mostly participant observation conducted during the author's lengthy trips with the informants through the tundra. The first-hand experience obtained from participant observation was supported by ethnographic interviews with the informants.

Pre-GPS Navigation Methods

In order to understand the process and effects of the GPS introduction it is important to take into account the details of spatial cognition of the tundra nomads as well as the navigation[2] methods they traditionally used before the introduction

1 *Orientation in, Perception and Utilisation of Space on the Edge of the Arctic: Nomads and Sedentary People in Northwest Siberia* (2005–08) and *New Technologies in the Tundra: High-Tech Equipment, Perception of Space and Spatial Orientation of Nomadic and Settled Populations of the Russian Arctic* (2008–12). Both research projects were part of the Collaborative Research Centre *Difference and Integration: Interaction between Nomadic and Settled Forms of Life in the Civilizations of the Old World* (Sonderforschungsbereich 586) sponsored by the German Research Foundation (Deutsche Forschungsgemeinschaft, DFG).

2 Tim Ingold (2000) finds it useful to differentiate between spatial navigation and wayfinding: in his opinion, the term 'navigation' should be reserved for instances of finding one's way by inference from a mental or artificial map, while the term 'wayfinding' should be applied in instances when the traveller follows routes that he or she remembers visually or in another perceptual modality. In my opinion, this differentiation makes little practical

of GPS. Both still existed in the Taz Tundra when the empirical material for this study was collected and, therefore, they could be directly observed and described (Istomin and Dwyer 2009).

The material collected among the nomadic and semi-nomadic inhabitants of the Taz Tundra seems to show that their spatial cognition is mostly based on memorizing geometrical relations between landmarks in terms of direction and distance between them rather than on the visual memorization of routes. In other words, the tundra dwellers have survey knowledge or a mental map of the territory they inhabit, rather than the routes or network knowledge of it (see Hart and Moore 1973, Shemyakin 1962, Siegel and White 1975 for analysis of this distinction). To put it simply, the tundra dwellers know directions rather than routes: they know in which direction one should start from one 'known place' to reach another, but they usually do not memorize the views or vistas one encounters during the journey, as many city dwellers do to 'learn a route' (Delvin 1976). During the travels with nomadic reindeer herders, the author witnessed several cases when herders lost, or believed they had lost, the sense of direction during the journey. As conversations with them revealed, they did not know (or at least were uncertain) where they actually were until some 'known place' was reached. Even if the direction was not lost (in the majority of the observed journeys), the herders in most cases insisted that they neither remembered the general view (vista) as it opened up from the sledge, nor any of its details (such as hills or bushes) on the way from one 'known place' to another. Routes are not memorized, but rather inferred from the knowledge of Euclidian relations between different places and it is only in the view of a 'known place' that the correctness of this inference can be ascertained (Istomin and Dwyer 2009). This emphasis on the abstract knowledge of directions and distances rather than on visually memorizing routes seems to be justified: it allows the Nenets herders and Nenets fishermen to navigate the tundra in poor visibility conditions, for example in darkness (which is especially important during the polar nights) or in mist, which is a rather frequent phenomenon in summer and autumn. What is perhaps even more important, this kind of space-knowing allows the tundra inhabitants to infer and follow a novel route: based on his or her knowledge of the directions from point A to point B and from point A to point C, a herder or fisherman can calculate the direction from point B to point C and find his way between these points even if he or she has never travelled between them before.

sense. Indeed, as I argued in detail elsewhere (Istomin and Dwyer 2009), spatial cognition of any person is likely to be based on both mental map(s) and route knowledge. Although the relative roles these two types of spatial knowledge play in spatial navigation can differ between individuals and cultures, many (if not most) instances of finding one's way can be based on inference from a mental (or artificial) map and route following, both of which can be applied either simultaneously or at different stages of travelling. Therefore, the term 'navigation' is used in this chapter to refer to the practice of finding one's way irrespectively of the type of psychological mechanism involved.

However, successful navigation based on the knowledge of directions and distances depends, on the one hand, on the ability to remember the directions from each 'known place' to other places correctly, and, on the other, to maintain the direction of travelling (e.g., not to lose course) between the 'known places'. The former is achieved mostly by aligning these directions with different natural features on and around the 'known place' (e.g., 'in order to go place X from here, I should start in the direction of that little hill nearby'). As for the latter, the inhabitants of the Taz Tundra have two basic methods of keeping direction. The first method, used mainly in summer, spring and autumn, relies on wind direction. When taking a course from a known place, a sledge-driver takes a note of the direction from which the wind is blowing on his body ('catches the wind', as the herders call this). During the course of further travel, the herder always tries to guide the sledge in such a way as to keep this direction constant. This enables him to keep to the course taken unless the direction of the wind changes – an event which reindeer herders consider the main reason for getting lost. The second method, used in winter, is based on *zastrugas* – sharp grooves or ridges formed on a snow surface in the tundra by a blowing wind. Their orientation is parallel to the prevailing winds. Starting a trip, a reindeer herder touches the snow surface passing under his sledge with his feet. This allows him to feel how the *zastrugas* hit his feet as the sledge crosses them, and, therefore, to feel the angle under which they are crossed. During the trip, the herder checks whether this angle has remained constant by putting his leg away from the moving sledge and touching the surface from time to time.

Navigation based on these traditional methods demands specific knowledge and skills. In order to find his or her way effectively, one should first of all (visually) memorize an extensive network of 'known places' as well as directions and distances between them. One should also master specific techniques such as 'catching the wind' and taking notice of the directions of *zastrugas*. It is important to stress that such knowledge and skills are not something the young tundra dwellers study by intentionally devoting their time and effort. Rather, these methods are learned through experience by travelling with elders. During these travels, through dealing with the situations and conditions in which navigation methods are practiced, each new generation of tundra dwellers experiences what Tim Ingold calls a 'guided discovery' (Ingold 2000) of these methods. Failure to learn these methods can result from the lack of experience rather than from the conscious refusal of the young generation to devote their time and effort.

However, even if these skills are mastered, navigation can still fail for a whole range of specific conditions. Thus, the described skills cannot be used during snowstorms, when 'known places' can be easily missed or misjudged due to poor visibility and catching the wind is difficult if not impossible. I believe that these limitations are very important for understanding the process of GPS introduction.

Adoption of GPS

In the Taz Tundra, the first GPS devices were adopted by the fishermen already in 2007, just a few months after their legal sale had been allowed. The most interesting fact about this early stage of the GPS introduction is that not a single device (as far as I was able to discern) was acquired for navigation. First GPS users purchased the device for one purpose: to efficiently conduct illegal or semi-legal fishing in the protected delta of the river Taz. This area is considered a highly important spawning ground for valuable species of sea fish. Moreover, nets with small meshes, which are very effective in fishing, are prohibited all over the region. The whole delta and most of the adjoining streams are regularly patrolled by officials using motorboats in summer and snowmobiles in winter. The patrols destroy illegal nets and fine their owners if they manage to identify them. Still, the fishermen have a long tradition of using illegal nets and exploiting protected territories. In order to do this, in summer the fishermen use nets with heavy sinkers and without floats – these nets, when cast, are completely covered by water and, therefore, are invisible for the patrols. In winter, the holes in ice used to cast the net are concealed by snow. However, to find and check the nets themselves, the fishermen either attached a single small float in summer or marked their position with a small stick or stone put on the ice in winter. Both marks could of course reveal the nets to the patrols. After GPS became available, fishermen quickly recognized that the device provided them with a new and extremely efficient way of concealing the nets. Now, the position of a net could simply be marked with a GPS device and later retrieved, remaining completely untraceable for the patrols. As one of the fishermen stated, GPS for him was a 'device to make an invisible float'.

This new option for illegal fishing resulted in an extremely fast spread of GPS devices among the fishermen. During the summer of 2009, I was told that for a fisherman GPS was considered as important a tool as a boat or a net. According to a rough estimate, most GPS devices possessed by reindeer herders and semi-nomadic fishermen to this day are still used for precisely this purpose as well as for marking hidden storages of goods and equipment rather than for navigation.

It was only during the last two years (2009–11) that the new technology has spread among reindeer herders and actually started to be used for navigation purposes. According to information obtained from the reindeer herders, their first encounters with GPS, which they still remember quite vividly, occurred during the exchange of winter visits with the fishermen. As mentioned earlier, the reindeer-herding nomads and the fishermen maintain regular contact and this is of great importance for both groups. Apart from relying on the herders for pasturing the reindeer herds some of them own, the fishermen also rely on the herders as suppliers of reindeer hides, which is the only material to make traditional fur clothes and cover nomadic tents (*chums*) in winter. In turn, the reindeer herders rely on the fishermen as suppliers of frozen fish, which is the main product to be sold or bartered in the village to obtain the goods they need. In addition, most of the representatives of each group have close relatives in the other group. It is important to note here that since the fishermen spend most of their time within the Taz river delta, their familiarity with the rest of the territory is rather limited. Therefore, the long-standing practice has been that the fishermen relied on the lead by reindeer herders when visiting distant camps, while the herders used the help of the fishermen when crossing the river delta. It was in the context of these visits that the first usage of GPS to assist in the actual navigation process occurred. The following account of a reindeer herder vividly describes the reasons which led him to purchase a GPS:

> Of course I have seen GPS among the fishermen for several years already. I thought: 'an interesting device, but useless in the tundra: we do not need to hide nets, do we?' But then a thing happened: once I visited Grishka (a fisherman) and we came to my chum together. He spent some time in my chum and then Grishka said he was going to go back home. I said: 'well, I will send my son to accompany you.' 'No need, I will manage myself.' 'How will you find your way?' 'I have a companion.' And he showed me a GPS and how it can be used. I thought: 'Good, but still I do not need that – I can manage in the tundra without the machine.' Then, the same year, another fisherman came and stayed in my chum for a while. Then he said: 'I will go home.' There was a snowstorm that day, one could not see anything, one could not catch a wind. I said 'how will you go, there is a snowstorm, [even] I would not find the way!' 'You cannot, but I can' and again, showed me a GPS. Then I realized – the thing was really useful! How many times I had to sleep in the tundra for several days because of a snowstorm! With this thing, I could always be at home by night. And I said: 'keep your fish and next time bring me the thing!' So he did. (Interview with Andrei Tyseda, Taz Tundra, February 2010)

As can be seen from this quotation, the situations in which the fishermen and the reindeer herders first came to rely on GPS for navigation are rather different. In the case of the fishermen, this was the need to find their way in the territory they had limited experience with due to the characteristics of their way of life. In the case of the herders, on the other hand, this was the situation of finding their way on the territory they know, but in conditions that hampered the application of traditional navigation methods. However, these two situations had an important thing in common: they both illustrate the objective limitations of traditional navigation methods. These limitations have always been acknowledged and they represent well-recognized problems, for which, however, no satisfactory solution existed. Therefore, GPS has now been adopted as a tool relevant for spatial cognition and navigation. However, it has not replaced the traditional spatial cognition and navigation practices, but rather complemented them. During my fieldwork in the Taz Tundra in the winter of 2009–10, I observed a growing number of reindeer herders carrying GPS devices on their trips. These devices, however, contained very few location marks in their memory: the home camps of the herders, several other camps, the village and several huts and gas drilling towers. According to the herders, they carry the devices 'just in case', to find their way home, should a sudden snowstorm occur. During the trips I joined, the devices were normally switched off all the way until the destination was reached. In fact, herders remarked, it was bad for the batteries to keep the devices running as they would quickly discharge due to the cold and as a result the device would fail, should the need to use it arise. GPS was 'the eye for the snowstorm', as one informant described it: 'with it I know that I will not sleep in the snow, whatever happens'.

It could be argued that the next stage of the GPS introduction in the Taz Tundra will consist in its gradual diffusion and replacement of the traditional system of spatial cognition and navigation. This stage has not yet been reached, but some of its precursors can already be observed among youths and a further study of their generation can shed some light on how the transition is taking place. Already during the winter of 2009–2010, I observed two young reindeer herders who relied on GPS more than others. More specifically, they did not only carry the devices with them during their snowmobile trips in the tundra, but also switched them on every now and then during short stops on the way to check the direction of their travelling. Unfortunately, only one of these herders could be interviewed in depth – the young Nikolai, who also gave me his GPS device for closer inspection. It was noticeable that his device stored more location marks than the devices of other reindeer herders inspected during the fieldwork.

Nikolai insisted that he knows the land – that is, the spatial distribution of 'known places' in the form of directions and distances between them – as required for the traditional navigation practices. His statement was supported by his ability to point to different landmarks with his finger and report approximate distances between them. In fact, the accuracy of his pointing, which I checked with a GPS, did not differ significantly from that of other reindeer herders of his age investigated during the fieldwork. This, as well as the fact that he had obtained the GPS device relatively recently and that he was able to find his way in the tundra before that, suggests that he had enough mastery of traditional navigation methods. His reasons for using the GPS device during his travels were the following:

> It is simply to be calm, you know. Without a GPS, you take a direction and then you do not know if you are following the right course until you reach the next known place. Even if you are sure that the direction was right when you departed, you always start to worry during the trip: am I still following the right course? Have I not accidentally changed the direction? Is the wind still the same or it has changed a little? Did I indeed take the right direction in the first place? And you have to worry like this until you come to a place you know. Then you are calm for a moment, but then all this starts again. I cannot stand it, you know. Old people have got used to that, but they have travelled all their life! I would rather take a look at a GPS and remain calm. If the GPS fails, I can still make it, but as long as it works, why should I have additional worries? (Interview with Nikolai Tyseda, Tazovskii town, February 2010)

This statement reminds me strongly of another account I once recorded from an old and experienced reindeer herder:

> The main thing you need [in order not to get lost] is to be brave and to go ahead. It is not easy. When you travel in the tundra, you always think 'have I taken the right direction?' and 'have I not missed the place I am going to?' Everyone has these fears, especially if you believe that you should have already reached a place but you cannot see any sign of it around, these fears become really strong. Now, you should not surrender to these fears. You should be brave! It is not easy, especially when you are alone in the darkness. You can think, for example, 'I have probably gone too far to the left, I should go a little bit to the right of the course I am taking now'. You can even eventually become completely sure about this, especially if you do not see the place when you think you should already be there. Still, you should not change the course. If you keep on the same course, you will eventually come somewhere, maybe not to the place you wanted, but still to a place you know. However, if you indeed change course at least once, you will get lost; because a little bit later you will start to believe that the new course is also wrong and you will change it again, then again. If you start changing course you will be unable to stop, believe me, nobody can. Then you will start to go in circles until your reindeer drop down and after that you will walk in circles. All the people who have become lost in the tundra and died did so because they were not brave enough and surrendered to their fears. (Interview with Nemechi Tyseda, Taz Tundra, June 2006)

Therefore, it can be argued that what actually makes the younger reindeer herders prefer GPS to the traditional orientation methods is not their unwillingness to invest their time and effort in learning them. It seems to me that such an investment does not result from a conscious decision anyway. Rather the young herders refuse to pay what can be named 'emotional' costs of using the traditional navigation practices. These costs are always present. Indeed, it can be argued that experienced reindeer herders stay home during a snowstorm, or stop and sleep in the snow if the snowstorm catches them during a journey, because of their perceived uncertainty that their traditional orientation methods would work in such situations. This is exactly the kind of situation for which the older reindeer herders acquire their first GPS devices. For the younger generation, on the other hand, the range of situations in which they feel uncertain is much wider – simply due to the lack of experience. For the same reason, the uncertainty related to the traditional navigation methods in their case is bigger than that associated with the GPS navigation (for example that the battery in the GPS would discharge). Therefore, they 'surrender to their fears', especially since the introduction of GPS has made the cost of this surrender low – some (electric) power in the batteries instead of one's own life. It seems to me that this can be the backdoor through which the 'satellite culture' (Aporta and Higgs 2005) penetrates the spatial cognition of the tundra nomads. Indeed, the limited experience of the traditional navigation practices created in this way naturally leads to the disappearance of these practices. It can be argued that the next generation of reindeer herders, following Nikolai's generation, will already be unable to learn the traditional navigation methods.

Discussion

According to the accounts given in the previous section, the history of the GPS adoption by Taz Nenets can be divided into three basic phases (Figure 10.1).

Phase of adoption	The way innovation is used	Impact of the innovation on traditional practices	Factors causing the transition to the phase
Original appropriation	The innovation is used to solve relatively novel tasks, for which there are no traditional solutions.	New and traditional practices exist side by side, each in its own sphere. The influence of the new practices on the traditional ones is minimal.	Mostly cultural factors (sociocultural causation)
Supplementary use	Both the innovation and the traditional practices are used for the same tasks. The use of the innovation is restricted to situations, in which the traditional practices are less effective due to their inherent limitations.	The new practice starts to impact the traditional ones, but does not replace them.	Both cultural and technological factors
Replacement	The practices based on the use of the innovation replace the traditional practices.	The traditional practices are replaced by the new ones. A chain of often important cultural and social transformations follows.	Mostly techno-economic factors (techno-economic causation)

Figure 10.1 Summary of the three-phase model of adopting a technological innovation

The first phase of adoption can be named the phase of original appropriation. This was the phase when GPS devices were mostly used to create an 'invisible float'. The main characteristic of this phase is that the innovation was used for purposes which were relatively novel and for which no traditional practices existed. Notably, these were not the purposes the innovation (GPS devices) had been designed for, i.e., spatial navigation. Traditional practices existed for this purpose and were relied upon. Since the diffusion of the innovation did not interfere with these traditional practices, its impact on traditional culture and way of life was minimal.

The second phase of adoption can be named the phase of supplementary use. It is the phase most Taz Nenets are currently experiencing. In contrast to the previous phase, the innovation is used for the same purposes, for which traditional practices had been used before. However, its use is limited to situations in which the traditional practices are not effective due to their inherent limitations. It is in this phase that the innovation started to interfere with the traditional practices and, therefore, to exert an influence on them. Still, the impact of the innovation on traditional cultural and social institutions is hardly significant.

The third phase of adoption, of which only precursors are currently observable among the Nenets, can be named the phase of replacement. During this phase, the range of situations in which the innovation is relied upon extends to the complete replacement of the traditional practices.

These three phases, therefore, differ in the degree of interference that the novel practices have on the traditional ones. They also differ with regards to the impact the technical innovation has upon culture and society. Importantly, the factors responsible for the initiation of each phase, i.e., the type of causation involved (techno-economic *vs* sociocultural, see the beginning of the chapter), also differ. Thus, I argue it was mostly 'sociocultural causation' which initiated the phase of original appropriation. Indeed, although GPS constituted an effective way of navigation that blended well with the local ecological and economic conditions, it was not appropriated for navigational proposes as the 'techno-economic causation' would predict. Rather, GPS was appropriated for tasks which were relatively new for the local population. Traditional means to deal with these tasks did not exist. Therefore, it was the new condition and the need to adapt to it (the sociocultural factor) rather than the effectiveness of the technology itself (the techno-economic factor), which determined how the technology was to be endorsed. At the same time, the transition from the phase of supplementary use to that of replacement can be better understood as involving techno-economic causation. Indeed, the scope of situations in which the new technology was relied upon broadens exactly because this technology is more effective (in reducing anxiety, for example). The transition from the phase of initial appropriation to that of supplementary use would depend on both the specific limitations of the traditional navigation practices (e.g., cultural factors) and the ability of GPS to compensate for them (e.g., technological factors).

Conclusion

The material presented and discussed in this chapter allows the proposal of a new approach to analysing the adoption process of technical innovation in specific cultures, the factors influencing this process and its cultural and social consequences. The basic idea of this approach is that adopting an innovation is not a unitary and coherent process. Rather, it can consist of several phases, which are caused by different sets of factors and have different consequences.

In my opinion, this approach represents a powerful way to juxtapose the sociocultural and techno-economic causation in the process of adopting technological innovations. Furthermore, it can explain why some innovations have a more pronounced transformative effect on culture and society than others and why some innovations affect one culture more than others. Indeed, it can be argued that the effect of a particular technological innovation depends on how far its adoption can proceed in a particular culture under the given social, cultural, economic or ecological factors. Indeed, the process of endorsing an innovation can be halted in the phase of initial appropriation if there are no limitations in the traditional practices that this innovation can compensate for. It seems to me that this is exactly why radio receivers did not have any effects on the culture and society of reindeer nomads (Pelto 1987): although they were handy in solving a relatively new problem in nomadic culture, namely that of getting information about society at large, it could not compensate for the traditional practices of long-distance communication inside the nomadic society itself, even if these practices had obvious limitations. Telephones, on the other hand, could compensate for these limitations and the effects of their introduction have been much more pronounced (Pelto 1987, Stammler 2009). Furthermore, under certain social and economic conditions the process of an innovation's adoption can be stopped in the phase of supplementary use. Thus, among the Taz Nenets, snowmobiles could not completely replace reindeer sledges in winter due to the limited supply and the high price of fuel. This effectively limits the use of snowmobiles in herding operations and explains why nothing comparable to the snowmobile revolution described by Pelto (1987) in northern Finland occurred in the region.

Further research is needed to validate this approach. Especially its applicability in other cultural contexts needs to be investigated.

References

Aporta, C. and Higgs, E. 2005. Satellite culture: Global positioning systems, Inuit wayfinding, and the need for a new account of technology. *Current Anthropology*, 46(5), 729–53.

Bateson, G. 1967 [1935]. Culture contact and schismogenesis, in *Beyond the Frontier: Social Processes and Cultural Change*, edited by P. Bohannan and F. Plog. New York: Natural History Press, 87–198.

Bateson, G. 2000 [1972]. *Steps to an Ecology of Mind*. Chicago, IL: University of Chicago Press.

Bauchspies, W.K., Croissant, J.L. and Restivo, S. 2005. *Science, Technology, and Society: A Sociological Approach*. Oxford: Blackwell.

Bloch, M. 1966. *French Rural History: An Essay on Its Basic Characteristics*. Berkeley, CA: University of California Press.

Delvin, A.S. 1976. The 'small town' cognitive map: Adjusting to a new environment, in *Environmental Knowing: Theories, Research and Methods*, edited by G.T. Moore and R.G. Golledge. Stroudsburg, PA: Dowden, Hutchinson and Ross, 58–66.

Donahoe, B. and Istomin, K.V. 2007. Izmenenie praktiki regulirovaniia dostupa k resursam u nekotorykh olenevodcheskikh narodov Sibiri: Popytka teoreticheskogo obobsheniia [The change of tenuriality among some Siberian reindeer herding peoples: Attempting a theoretical generalization], in *Rasy i narody: Sovremennye etnicheskie i rasovye problemy*, vol. 33, edited by D.A. Funk. Moskva: Nauka, 128–63.

Dwyer, M.J. and Istomin, K.V. 2008. Theories of nomadic movement: A new theoretical approach for understanding the movement decisions of Nenets and Komi reindeer herders. *Human Ecology*, 36, 521–33.

Engels, F. 1972 [1884]. *The Origin of the Family, Private Property and the State*. Reissue. London: Lawrence & Wishart.

Harris, M. 2001 [1968]. *The Rise of Anthropological Theory*. Walnut Creek, CA: AltaMira Press.

Hart, R.A. and Moore, G.T. 1973. The development of spatial cognition: A review, in *Image and Environment: Cognitive Mapping and Spatial Behavior*, edited by R.M. Downs and D. Stea. Chicago: Aldine, 246–88.

Hughes, T.P. 2006. Culture and innovation, in *Cultures of Technology and the Quest for Innovation*, edited by H. Nowotny. New York: Berghahn, 27–38.

Ingold, T. 2000. *The Perception of the Environment: Essays on Livelihood, Dwelling and Skill*. London: Routledge.

Istomin, K.V. and Dwyer, M.J. 2009. Finding the way. *Current Anthropology*, 50(1), 29–49.

Istomin, K.V. and Dwyer, M.J. 2010. Dynamic mutual adaptation: Human-animal interaction in Reindeer Herding Pastoralism. *Human Ecology*, 38(5), 613–23.

Kroeber, A.L. 1952. *The Nature of Culture*. Chicago, IL: University of Chicago Press.

Pelto, P.J. 1987 [1972]. *The Snowmobile Revolution: Technology and Social Change in the Arctic*. Prospect Heights, IL: Waveland Press.

Sahlins, M.D. 2000. *Culture in Practice*. New York: Zone Books.

Shemyakin, F.N. 1962. General problems of orientation in space and space representations, vol. 1, in *Psychological Science in the U.S.S.R.*, edited by B.G. Anan'ev. Washington: Office of Technical Services, 186–255.

Siegel, A.W. and White, S.H. 1975. The development of spatial representations of large-scale environments, in *Advances in Child Development and Behavior*, vol. 10, edited by H.W. Reese. New York: Academic Press, 9–55.

Stammler, F. 2009. Mobile phone revolution in the Tundra? Technological change among Russian reindeer nomads. *Folklore*, 41, 47–78 [Online]. Available at: http://www.folklore.ee/folklore/vol41/stammler.pdf [accessed: 18 November 2012].

White, L.A. 1949. *The Science of Culture*. New York: Grove Press.

Chapter 11
Narratives of Adaptation and Innovation: Ways of Being Mobile and Mobile Technologies among Reindeer Nomads in the Russian Arctic

Florian Stammler

Yamal means 'land's end' in Nenets. It is the name of a West-Siberian peninsula, where I conducted long-term anthropological research that provides evidence for this chapter. This name is not given by outsiders but by the Nenets people themselves. Such a designation of the territory points to a perception of remoteness even by the very inhabitants of this peninsula that these days has become more famous for its gas reserves and their importance for European supplies (BBC 2012). On its 149,000 square kilometres the Yamal Peninsula hosts approximately 5,000 full-time nomads with yearly migration routes of between 100 and 1,300 kilometres in length (one way), most typically in a north-south return journey. The nomads manage their reindeer in herds of various sizes, depending on the season and on personal wealth, between 20 and 5,000 animals.[1]

The area as well as its human and animal inhabitants have witnessed and taken part in several big waves of technological change, induced either by state ideology during the Soviet Union, by the slow trickling down and diffusion of technological innovations, and more recently and radically by the advent of the gas exploration and extraction industry on the peninsula (Stammler 2011). I propose in this chapter that in addition to analysing these big waves of development, we can understand the implications of technological change for the people best through narratives, told as a part of people's life histories.

The main objective of this chapter is to make this link between what I call narratives of adaptation and innovation that reveal to us how Nenets nomads participate in technological change, and the analysis of how their involvement in these waves of change also alters the very notion of their homeland Yamal,

1 Speakers of Nenets, this Samoyedic language, belong to the Uralic linguistic group. According to the Russian census of 2010 there are approximately 45,000 Nenets, of which almost 30,000 live in the Yamal-Nenets Okrug, almost half of which lead a nomadic way of life. For more details on migration routes and the determinants of grazing land use, see Stammler (2005: ch. 6) and Behnke et al. (2011: 158–65).

the end of the world. These ideas become important for our further theoretical understanding of information flows and flows of nomadic movement as mutually fruitful conversations in a field of increased global integration. Understanding these conversations between people, animals, the land and space contributes to the discussion of whether perceptual processes are based on experience or rather mental representation (or both). I shall explore how the apparent opposition between these two ways of seeing people perceive the world (Ingold *versus* Howes 2011) can be resolved using an anthropological analysis of technological change in a nomadic livelihood.

Theorizing Technologic Change and Innovation

Ethnographies of material culture have focused on describing tools and technologies used within a certain society that were thought to be essential markers of their livelihood. Among the Yamal-Nenets and other Arctic nomads, such markers are the conical tent (*chum*, *mya*, Anderson 2007), the sledge (*narta*, *khan*) for moving through the land, and the male and female clothing that permits surviving in the cold during long migrations. On the other hand, much of the material culture of Nenets nomads today is the result of technological innovations imported from neighbours and then creatively adapted to fit the nomadic lifestyle of the reindeer pastoralists. The import of wheat (bread) and iron into the nomadic culture of West Siberia happened much longer ago than can be remembered today (Gusev 2010), but talking to elders we can hear about the influence of more recent innovations, such as the iron stove in the *chum*, the introduction of black tea as a beverage, rubber boots as footwear, and the replacement of birch bark by canvas as a cover for the summer tent. The position of these innovations in individual lives helps us understand the more general trajectory of narratives of change in the society.

While every narrative is individual for its context in somebody's biography, there are certain general features and trends that can be observed: Firstly, change is conceptualized by many people themselves as a movement from one stable stage to another – i.e., how life was before and after the adoption of an innovation, even though the reindeer herders' narratives of innovation often focus on the process of change. This points to a certain divergence between how people *experience* change – usually as a flow in a constantly changing nomadic lifestyle – and how they *think* and *talk* about it, using categories of time to delineate certain stable stages that during the actual experience did not exist. I argue that these stable stages are cognitively constructed *after* some experience of change. In the Russian northern context, education in a Marxist-Leninist ideology where human development goes through a series of stable evolutionary stages certainly played a role in shaping people's processing and categorizing their own experiences of change (Yurchak 2006).

Secondly, there is a general tendency in Nenets society to see the influence of innovations positively and explore them with great curiosity. This may also stem from the exposure to Soviet ideologies of progress and social evolution, i.e., that those using old technologies are considered backward and primitive. However, I argue that this openness to innovation also stems from a high adaptability and flexibility of this nomadic society to accommodate changes in any sphere of life (Stammler 2005: ch. 9).

Thirdly, at the same time, negative consequences of an innovation are often conceptualized in terms of losing skills and independence in older people's narratives as well as some social science accounts (Pelto 1987 [1972], Harrison, 2007), even though ethnographic evidence shows that people can revive old subsistence skills if needed, which we see in Ziker's work (2002) among the Dolgan and Nganasan or the Eveny in Vitebsky's book (2005: 380 *et passim*)

This leads us to investigate a certain discrepancy between data generated through oral history research where people use their learnt categorizations, and those stemming from participant observation living with the people in the process of enacting and driving the change. In this context I favour the use of these two verbs *enact* and *drive* rather than *happen* and *occur* to continue a discussion from Ingold's recent work (2011: 143) because the former emphasize people's agency in these processes. Is this discrepancy one between dwelling and experience on the one hand, and cognition and thinking on the other? Field research using both oral history and participant observation methods suggests that conceptualizing memory in stable states is largely the result of a Soviet-modernist education process among my fieldwork partners. Ingold (2011: 142–3) attributes this classificatory form of knowledge (i.e., including memory) to the influence of the genealogical model that we have so much internalized that we do not notice the context-boundedness of our experiences and the ideas resulting from those. Just like the spatial and temporal fluidity of nomadic life was considered backwards in the twentieth century, so was the steady contextual experience of change and the nomads' perception of life-cycles. In other words, seeing change in terms of a transition from one stable stage to another, best on the ladder of a Marxist evolutionary model, was the desired official interpretation in the Soviet Union. Khlinovskaya-Rockhill (2010) shows how these ideas still live on. Also beyond her immediate cases, such a worldview structures people's ways of making sense of changes in their life.

Processes of adopting innovations in society have been most prominently theorized by Rogers (2003 [1962]), who with his theory reminded social scientists of the importance of a diffusionist perspective if we want to understand how societies incorporate innovations. Rogers describes such diffusion as 'the process in which an innovation is communicated through certain channels over time among the members of a social system' (2003: 5). So we see that the diffusion of an innovation is a flow of communication – which is in turn in Nenets reindeer nomadism tightly connected to flows of nomadic movement. The diffusion process goes according to Rogers through (a) *knowledge* about an innovation, via (b) *persuasion*, (c) *decision making*, (d) *implementation* and (e) *confirmation* (Rogers 2003: 169–70). The people driving this process are according to Rogers in five different categories: 1) the *innovators* who have the means and contacts to acquire the innovation and are ready to take the risks of the little-known; 2) the *early adopters* who see the technology first among innovators and then become opinion-leaders in their society to incorporate it; 3) the *early majority* who follows the opinion leaders and gradually take the innovation over; 4) the *late majority* follows what Rogers calls the 'average members' of a society. These people are more sceptical but are eventually convinced; 5) the *laggards* are – as the name says – those who lag behind and care more about the traditions, being sceptical about the change that adopting an innovation involves (Rogers 2003: 282–4).

Such a categorization is definitely vulnerable to many valid criticisms, in particular as to Rogers's rather careless connection of social status, financial means and leadership roles to his categories mentioned above. I would also emphasize that his classification of five stages, from (a) to (e), is his attempt to build a model for explaining *processes*. In lived experience we can see that these are not stable stages or a steps in a ladder that societies jump up – rather they are crutches for us scholars to understand processes as they occur which we, with our educational background, may not understand otherwise. Indeed Rogers (2003: 195) himself added that 'we should think of stages as a social construction, a mental framework that we have created and generally agreed to'.

Stripped of these questionable links, his theoretical description of these processes still serves as a generally applicable explanation model of how individuals in a society deal with innovations. Using Rogers's model, we can trace the diffusion waves rereading Pelto's classical ethnography (1987 [1972]) of the introduction of snowmobiles among Sami reindeer herders. There, it was a teacher who first found out about snowmobiles and bought one for recreational use. This is Rogers's innovator (1). Reindeer herders found out about this (stage (a): knowledge) but without using it yet. After that they were (b) persuaded by the potential usefulness of the innovation – the snowmobile – in their life. In deciding to buy a snowmobile for their own work (c), the first herders were early adopters (2) according to Rogers's terminology. When this became a mass phenomenon because of the early adopters as opinion-leaders in the society, Rogers's early majority (3) of the society acquired these machines, they implemented the innovation into their own lifestyle (d). Gradually, even those who could not afford snowmobiles started asking their neighbours to use them even though they were initially sceptical; they were Rogers's late majority (4). With this step, the innovation was confirmed (e) in the society and used to its full potential. Behind stay Rogers's 'laggards' (5), those who do not use a snowmobile.

As appealing as this explanation model may be for its simplicity, many of the fine grains of this process are missing in the general categories, e.g., the distinction between those who cannot afford to use the technology and those who do not want to use it. Pelto's ethnography of this technological diffusion has also been criticized for its coarseness: in fact, the changes he portrayed are not as universal and as negative as they may appear (see Stammler 2009: 49–54). More generally, we do not find out from Rogers's analysis how exactly the process of persuasion works, and how people make the decision to incorporate an innovation into their own lifestyle or not. Rogers (2003: 13–14, 259) distinguishes between the 'hardware' and the 'software' aspect of technological changes. In the case of the snowmobile, the hardware is the physical object, the tangible machine in its usefulness and shortcomings for performing particular tasks in practice – here for example collecting a reindeer herd. The software is the recognition of the potential ways in which the innovation should be adopted to local conditions to make performing the particular tasks required possible, a notion I shall link in the following with the idea of affordances (*sensu* Ingold 2011: 77–9).

For harmonizing both the hardware and software aspects of the technological change resulting in the innovation actually being implemented, I argue that a combination of cognitive processes and lived experience is necessary. Again ethnographic evidence shows us that the two – cognition and experience – are not mutually exclusive as was argued in fierce academic debates (Ingold *versus* Howes 2011). Ingold argues against any socially or culturally transmitted systems of representing the world around us and emphasizes the universality of perception as an individual and situational process in the environment based on practical experience. Howes argues for the existence of collectively shared ideas in society which cognitively influence the ways in which humans perceive the environment. In my reading both authors actually agree towards the end of the debate unintentionally: when Howes writes that 'one will find worlds of sense shaped by ways of sensing and ways of sensing configured by worlds of sense' (2011: 331), he refers in my reading to the same intertwining of cognition and experience as Tim Ingold, who argues that bodily sensation and mental representation belong together, adding that 'it is out of this interweaving that meanings emerge' (2011: 326). Istomin and Dwyer (2009) argue for the mutual complementarity of these views, while I would go further and suggest that they are both indispensable: human sensual perception is based on the experience of particular conditions and contexts during a process in which the affordances of technology are re-cognized. For example, the hardware of the snowmobile affords a use as a reindeer herding tool, even though this was not the reason it was invented or introduced to a reindeer herding community in the first place (Stammler 2009, Pelto 1987). Yet only through the process of seeing the snowmobile as a possible reindeer herding tool does the decision making and the subsequent implementation of the innovation happen. This seeing and decision making happens in the brain of a herder at some point. But it does not happen in a vacuum: it happens, for example, while a reindeer herder is going on a leisurely snowmobile ride with the teacher, i.e., during an actual experience, or while seeing the snowmobile being used otherwise. With this description I deliberately do not intend to speculate if the sensual experience or the cognitive process comes first. All I argue at this point is that both are not only complementary but necessary for the innovation to be incorporated in a particular livelihood.

If experience and cognition do not work together, and/or the hardware of the innovation does not meet local demands and conditions at a given time, the innovation will not be incorporated in the society. However, ultimately it is the potential users who decide if they find the ground fertile or not for the technology to be applied. In reindeer herding, we have many examples of how first the hardware of the technology needs to be changed or reinterpreted to become fit for the environmental and socio-cultural dimensions of the space where the technology is used every day: e.g., snowmobile-chassis being reinforced for heavy forest use, or fishing nets remoulded and recycled to become reindeer lassos (Stammler 2002), or the video and global positioning system (GPS) function of a mobile phone used for reporting an oil spill (seen in the Nenets Autonomous Okrug of Northwest Russia, Stammler 2009: 65), or GPS devices used by fishermen to place their nets under the water surface without marking them and thus to avoid trouble with the fishery inspectors (Istomin, this volume).

Drawing on the idea of affordances in particular contexts, I suggest that the analysis of technological change should start with looking at the essence of what a device offers. One crucial asset is versatility, and that is what has made personal communication technology (PCT) devices and mobile phones in particular so successful (Stammler 2009). The affordances of the device will determine how it influences a particular society, but as Caron and Caronia (2007) convincingly argued, not only technology transforms culture, but culture also transforms technology, precisely through the software aspect of an innovation: it can be seen differently by a reindeer nomad and an urban metropolitan teenager for one and the same device, e.g., a mobile phone.

Narratives and Technologies of Mobility

In a nomadic society, such as that of the Nenets reindeer herders and fishermen, every technology is mobile technology – if it is not, it is either not used or made mobile. Only innovations that can be used in a nomadic lifestyle are adopted. This basic precondition should be kept in mind for the following sections of this chapter, when the importance of larger narratives of innovation and adaptation shall be analysed. The snow*mobile* and the *mobile* phone are exemplary candidates of innovations which can become well embedded into a society of *mobile* pastoralists, but not the only ones, as shall be shown hereafter.

Iron Stove, Mobile Cooking and Heating

In Nenets society, long before those two mobile innovations just mentioned, the introduction of the iron stove to the *chum* (conical tent) made moving in winter and heating with ocean driftwood much more comfortable. By replacing the open camp fire in winter, mobile housing for the first time became comfortable even in harsh temperatures in areas with cold coastal storms. Thus, the adoption of the iron stove can be seen as the diffusion of an important technological innovation contributing to the Nenets nomadic mobility. As a result, the perception of the environment also changed because a stove needs to be fed with a specific kind of firewood to be efficient. So people started looking for larger pieces of wood to make longer-lasting fires in the stove. Before, the idea of using firewood as a source of heat was not at all important– firewood was only used for cooking, and appropriate clothes were there to protect from the cold, so there was no need for heating. Among many herders with whom I have stayed this idea is prevalent to the present day, so that stoves are mainly used for cooking, the heating aspect being a welcome side-effect but not the purpose of firing a stove.

Figure 11.1 Three types of mobile fuels: wood is used for the iron stove (right hand side); gas for the cooker at the back; and kerosene for the lamp above it

Source: Photograph by Florian Stammler

Obviously, heating and cooking are crucially important topics of nomadic survival in the Arctic, and the Nenets herders' general openness towards innovation and mobile technologies in this respect was demonstrated to me by a herding friend in spring 2011: 'Why do we still use firewood, this stone-age sort of fuel, in our tents? Under the ground we have the world's largest gas deposits. Why don't the companies provide us with efficient gas stoves that we can easily move around? Gas is lighter than wood as a fuel, stoves can be carried more easily, and less work for cooking [is needed]. This possibility should be looked at' (Yakov Vanuito, 14 April 2011). This shows how perceptions of the affordances of the land can change, and points to the cognitive processes at stake in this perception: in this example we do not only have the basic processes involved in obtaining and storing knowledge, but more elaborate processes in which knowledge, here about the gas resources under the pastures, is already available and now reason is applied for putting that knowledge to use. I would still argue that most probably this is related to the experience that Yakov had in seeing gas stoves used in the village. Without that experience he might not have extrapolated the use of gas instead of firewood as fuel for nomadic tents. On the other hand, in spite of all openness for innovations, we can still observe that older mobile technologies have not been completely replaced in this society, which formed one of the key arguments made in previous works (Stammler 2005, Beach and Stammler 2006): open fires are still commonly used during the snow-free season, and even though light from electric generators is nowadays more the rule than the exception, we still find kerosene lamps in every *chum* as well. This shows that people are 'adopters' and 'laggards' at the same time, to put it in Rogers's terms (2003).

Walk-Wear for Mobility: Rubber Boots

Another diffusion of a mobile technology can be observed with the introduction of rubber boots: I remember well how the reindeer herding veteran Pupta Yamal and his wife told me how before Second World War they crossed the rivers in spring when the ice had just broken up and placed fishing nets there. They did that barefoot, saving their sealskin fur waders from too much moisture and wear. Every time their legs would hurt from the freezing cold water. Talking to me in their old years, they were sure that their leg problems resulted from that. After the war, they got rubber boots, which quickly became the principle footwear among all tundra dwellers for the snow free season. Obviously, after the introduction of the rubber boot, nobody would walk barefoot anymore in the cold swamps that cover the permafrost of Yamal in summer.[2] Also, the bottom of a rubber boot is harder than that of a skin-boot, so the haptical contact of the foot with the ground is mediated by the boot. This is how the *hardware* of the rubber boot changed the process and feeling of walking – still a highly important mode of nomadic mobility in the area. However, the *hardware* of rubber boots was not suitable beyond the summer months, problems remaining in spring and autumn when the snow is wet. Rubber boots were not big enough to be worn on top of reindeer skin boots for moisture-protection. This is why the thin protective rubber boots used in chemical

2 These narratives of earlier technological change were condensed out of live history interviews made with elders on the northern Yamal Peninsula during fieldwork in 2001 and 2011.

Figure 11.2 The chemical protection rubber boots on top of the herder's skin boots

Source: Photograph by Florian Stammler

production and nuclear power plants came into use all over the tundra so fast after their introduction (Figure 11.2). Here again herders foresaw a use that was never planned by those who invented this sort of footwear. However, during my fieldwork in spring I used water-resistant boots made of sealskin myself, which were valued traditional footwear despite the diffusion of rubber boots. Sealskin boots were valued because during work outside in spring and autumn the feet do not sweat in them, while the chemical protection rubber boots on top of reindeer skin footwear inhibit evaporation and do not allow the feet to breathe.

Snowmobiles

Less than 40 years lie between the introduction of the rubber boot and the snowmobile in this field area. Both of them had the effect of changing the process and feeling of nomadic mobility. However, it seems the process of diffusion of the snowmobile did not happen as fast and as thoroughly as in the case of the rubber boot. I have outlined the reasons for this in detail elsewhere (Stammler 2009), but it is worth mentioning that Rogers's same explanation model of the diffusion process that worked well for Pelto's snowmobile ethnography among the Sámi does not work for the field in West Siberia. The reason is that the model assumes freedom of choice for using an innovation and sufficient available supplies for the new technology. Neither was it the case in the Soviet Arctic. Ideological, political and economic reasons prevented a broad diffusion of the snowmobile among reindeer nomads in Yamal. The Soviet Union clearly prioritized large-scale transport mechanization with helicopters, tractors and tracked vehicles (*vezdekhod*) for the tundra, whereas individualized, mechanized transport was seen more as a consumer good and, from the point of view of state planners, not desirable for the practice of reindeer herding. So buying a snowmobile within the state herding collective was either very hard or simply impossible. Then towards the end of the Soviet Union when it became possible, there was a shortage of spare parts and fuel to the extent that anthropologists working in neighbouring regions observed a return to subsistence practices that required a minimum of goods imported from outside the region altogether (Ziker 2002).

In addition, the physical conditions in much of Fennoscandia are more favourable for the diffusion of individual mechanized transport. In Siberia, distances between populated centres are much further than any snowmobile fuel tank can reach. Even today petrol stations are not available in all villages of the area. For example, I came in April 2011 to the camp of Mikhail Tusida to find out that he had left from the eastern side of the peninsula all the way to Kharasavei on the West coast to get petrol for his snowmobile. This meant he had to burn as much petrol for the return trip as he could bring home! 200 litres for the trip and 200 litres carried home. An airplane, helicopter or *vezdekhod* instead can go much longer with one fuel tank, so snowmobile technology is not such a good fit for this environment.

Nonetheless, in comparison to the early 2000s, there has been a sharp increase in the use of snowmobiles in the last five years, particularly among nomadic fishermen and herders with small herds and short migration routes. Moreover, more affluence in the nomadic society has enabled tundra people to buy more snowmobiles, even though many use them only for social activities such as bride-search, visiting friends and relatives, trips to town etc. This increase has been exclusively due to privately bought snowmobiles, whereas earlier some herding camps got them allocated collectively.[3] Although private herding and fishing in the area is on the rise, I would, however, not argue that this has led to an individualization of economic practices altogether because herding (maybe more than fishing) still requires a group of people to cooperate. In the furthest north, I saw herders doing the actual day-to-day herding work on a snowmobile in 2001 for the first time. By 2011 that had become more widespread. Young people – I heard comments by elders – have become lazier, they have to catch and train less draft animals than their fathers' generation, and they spend so much time fixing snowmobiles that there is no time for reindeer training any more. On the other hand, it would be overstated to say that economic relations have been reorganized in this case.

3 Camp is here used for the Russian *brigada* – a team of herders employed by one of the agricultural enterprises of the region.

I have shown elsewhere (Stammler 2009) why this is one of the differences between the West Siberian case and Pelto's (1987) analysis of snowmobile diffusion in Lapland. Even considering developments in the last three years we can see how an expensive imported snowmobile is considered a nice toy for bride search, but not a symbol of prestige and wealth. The principle indicator remains a big and beautiful reindeer herd – not a snowmobile, which usually has a life expectancy of just two years anyway. People even started to use old Soviet style snowmobiles in summer in the swamps for herding work (see also Klokov 2011). However there they wear out and break even faster, which is why this can never be a reliable mobility option for nomads with long migration routes. Therefore all the herders with bigger herds still do their long distance migrations using reindeer and sledges all year round.

In my previous work (Stammler 2009) I described how driving snowmobiles changes the perception of the land and snow. This is first and foremost connected with seeing the affordance of *driveability* on the earth surface. Driveability depends to a large part on surface hardness, snow texture, vegetation and terrain. A snowmobile driver experiences and evaluates all these factors differently from a reindeer sledge driver or a pedestrian. From this follows that on the one hand, mechanized transport does not afford the same intimacy of the knowledge that a reindeer driver has as he/she perceives through movement, the interaction between the human body, the pulling animals, the vegetation, the water and the land on which the movement happens. On the other hand, the set of skills and knowledge acquired through snowmobiling experience and other mechanized tundra transport generates a perception of the land that has its own specific quality not to be classified as more or less complicated. The snowmobile, particularly modern models, also gives a different experience of distance due to the increased speed. It is noticeable that, for example, the area young herders with good snowmobiles cover when searching a bride is bigger than the area for those searching for brides using reindeer transport.

As a result, the traveller gets to know the environment on the way in less detail, but gets a better idea of places more remote from his route leading to an increased awareness of the vastness of the land. On the other hand, not only in Yamal elders' narrative do we find a whole wealth of clues about the perception of space and place through the long travels on reindeer that the old people used to do. Anderson (2000:131) gives a vivid example, showing that the old people travelled so far before mechanized transport that younger people cannot imagine any more. Argounova-Low (2012) beautifully describes how formerly long travel routes have been abandoned physically but live on in people's memories (see Habeck, this volume, for more details on both reports). Therefore we cannot say that uniformally the snowmobile has increased people's awareness of the vastness of the land, nor has it always increased distances covered by people in their journeys as was argued before (Kemp 1971). Maybe on the contrary, travel activity has become restricted to the range of fuel tanks.

Mobile Phones

While earlier publications have emphasized the radical newness of the mobile phone as an innovation different from any other (Katz and Aakhus 2002, Stammler 2009), here the analysis shall be focused on the ways in which the introduction of the mobile phone fits into broader narratives of innovation and adaptation within the nomadic society. For this purpose I would suggest to look at PCT devices altogether rather than isolating the phone as a calling device. Nowadays phones include such a diversity of communication tools – text messages, MMS, photo-video shooting and sharing, GPS, internet, chatting, astronomical apps that one can use when looking at the stars, and other applications – that it is worth analysing the use of the different tools in one device together. I therefore argue that communicating with satellites in space using the GPS function of a phone and communicating per text message with distant friends are both ways of using PCT as conversations with one's environment.

Looked at in terms of mobility, the mobile phone like the snowmobile has changed the way people perceive distance, and their thoughts while being mobile. It has also changed the pace of nomadic migration for some, which can lead to an entirely new structuration of space into *off* and *on* regions. Caron and Caronia (2007) coined this 'on' and 'off' distinction for the generationally different acceptance of new PCT devices. In the following I show how 'on' and 'off' is useful in both ways: firstly, to distinguish space with and without network coverage, with all the implications this has for mobility and social relations, and secondly to characterize users of particular functions of PCT devices, as I shall show below. During my fieldwork in summer 2012, I saw tundra reindeer nomads for the first time enact what these authors (2007) described as the process of verbal performance through which people create culture: young herders standing on an elevated spot with mobile coverage used one of the internet-based new chatting services in Russia called *drug vokrug*[4] (friend[s being] around). Mostly these were friends of their own region, mates from school, but also partially from other corners of the country, where some boys had been doing their army service. While the recent diffusion of TV and films in the tundra goes across all generations and creates a new collective experience, the use of the extended functions of mobile phones is almost exclusively by people under 30 years old, an observation confirming Caron and Caronia's emphasis on generational differences.

More importantly, what we see in the Russian North is that *on* islands re-structure nomadic space. For example, MTS – one of the largest operators in Russia – stated in 2005 that the expenses for building a transmission tower only pay off if there are at least 1,000 potential customers in the catchment area.[5] That condition is met by very few indigenous nomadic groups, where one big herd

4 www.drugvokrug.org

5 Information by Vladislav Peskov, valid for the European Nenets Autonomous Okrug (see Stammler 2009: 62).

alone needs some 15 or 20 km around them, which means the neighbour's camp should be located no closer than that. Therefore, small islands of *on* regions exist mostly where there is a significant sedentary population, often connected to the extractive industry.

Herders and fishermen started classifying space along their nomadic migration routes in terms of 'on' and 'off' mobile coverage. For example, while migrating in summer 2012 with the Khudi brothers from Camp 4 from the Bovanenkovo deposit towards the Arctic Ocean, one of them – Evgenii – told me before the next migration of the camp: 'on the next camp site we will have moved beyond mobile coverage, and the next time that we will have [access to] a network will be only in a month's time'. When we returned to the *on* coverage area, Evgenii's nephews happily looked forward to going to the highest spot with the best signal to chat in the *drug vokrug* social network. In doing so, they bridge geographic distances that in the pre-mobile phone age had to be walked or reindeer-driven entirely during summer, or snowmobiled during snow-cover.

While Caron and Caronia (2007) see a liberation of constraints in time and space in the use of such communication technologies, I have also witnessed the opposite in the Russian tundra: the structuration of space in *on* and *off* areas creates spatial and temporal constraints, when people feel they have to walk half a day in bad Arctic weather just to reach the closest phone signal because for example their fiancée in the *on* area expects them to reply to a text message. On the one hand one could argue that the constraint comes from the lack of full mobile coverage everywhere – on the other hand nobody would expect any reply, were there not some network catchment area in the vicinity. So the principle availability of a network creates expectations leading to constraints in time and space on the daily work routine in a nomadic lifestyle. In both of these cases it is legitimate to speak of a dialectic relationship among objects and spaces: technologies define spaces (here: *on* and *off* areas perceived by herders) but are also defined and redefined by their location (here: in *off* areas mobiles are only used for games, photos, and videos) (Caron and Caronia 2007: 65).

Narrated and Teleported Journeys

The examples of the innovations mentioned above and some of the narratives connected to their introduction show us that the perception of the land, and the environment around it that some scholars like to call space (Massey 2005) and some refuse to (Ingold 2011: 141–2 and 145–55), is indeed strongly influenced by technological change. In my analysis, I shall therefore focus on how technology influences these flows of movement and communication at the heart of nomadic life. All of the innovations described above have influenced the ways in which people see, feel, experience but also think and reason about the land they live on.

The extent to which time and space in the tundra are related to the use of these innovations becomes clear from herders talking about their journeys. More and more we hear how many kilometres they travelled from one place to the other, because that is what is shown on the speedometer of a snowmobile (if it works properly). Increasingly GPS devices or the GPS function within phones are also used to measure distance between two points. Herders always enjoyed not only making journeys but also following other people's journeys sitting in the *chum* and talking about them. The availability of devices measuring distance has added another point of interest to such conversations. In particular, it was with the GPS that direct distances measured *po vozdukhu* (as the crow flies) first appeared in people's ideas. Many had not worked with maps before where they could see absolute distances. Therefore, in the conversations about journeys the difference between the kilometres measured by a snowmobile speedometer and a GPS is of interest. But even greater is the interest when a GPS device records the actual distance migrated on the ground – through a track log. This is something our research group did in previous projects (Forbes et al. 2009), but I have never seen herders themselves using GPS for recording distance.

As important narratives are as constituents for the perception of the land (Argounova-Low 2012), a nomadic lifestyle is more about a practice-oriented way of relating to one's environment, enacting and generating ways of knowing, perceiving and responding. Therefore it is crucial for people to physically experience their environment through travelling. Walking, reindeer sledging as well as travelling by snowmobile give such experience. Mobile phone communication has slightly changed this because here information travels without physical people. Information travels in a virtual way that before had been known mostly by shamans, whose soul travelled to exchange information with spirits, with the physical body remaining where it is (Stammler 2009). Taking this argument further, we can see that this potentially diminishes the need for the physical movement of people across the land. In summer 2012 I witnessed reindeer herders' discontent about having to travel to some neighbour for searching reindeer just because his camp was out of mobile phone reach: 'would he put his camp within the coverage zone, we could just call and find out if our reindeer are in his herd. Now we need to go there'. That complaint was because of the additional work that travelling involves: catching and harnessing reindeer, travelling hours and searching for the animals. In early August 2012, the leader of Camp 4, where I was staying, decided to forbid his young herders to take part in the summer reindeer races and sports competitions organized by the neighbouring Camp 8. The reason was that work had to be done for searching reindeer from the neighbours' camps. The young herders complained that they could have gone to the festivities if the neighbours had a mobile connection. When I said that even so, the reindeer would have to be fetched from the neighbours and so physical travel would still be necessary, they replied that this could have been done at a later stage as long as the camp leader knew where the animals are.

This leads us closer to analysing the connection of knowing the land and physical travel, and the way information flows through space. Obviously the mobile phone has radically changed this, but for indigenous northerners the flow of information before physical travel is not entirely new. Vitebsky relates the exceptional sensitivity of herders towards any mood, movement or unspoken thoughts to the quietness of the northern environment that he calls 'the long silences of the taiga' (2012: 439). The Eveny, with whom Vitebsky and his colleague Ulturgasheva work, have a name for this sensitivity called *djuluchen* (Ulturgasheva 2012: 46). *Djuluchen* is a way of sending one's intention of travelling and arriving at a destination ahead in time and space of one's physical travel like a shadow of one's own person. So through *djuluchen* hosts would already sense long before the arrival of guests who comes. I see this similarly to sending a text message to your destination announcing your arrival. Note that this is very different from calling, because the text message is not actually spoken word, it is more discrete and does not impose a loud noise interrupting somebody else's routine. All three, *djuluchen*, text messages and phone calls may be relevant for what Westerners call telepathy. Sheldrake's research shows us that telephony (be it for text messaging or calling) does not have to replace telepathy, but rather complements it: instead of telepathy decreasing because of the use of mobile phones, he found that there is telephone telepathy at stake above the chance levels when callers know who calls them before they receive the call (Sheldrake, Avraamides and Novák Mobifi 2009).

It should also be noted that the travel of people and flow of information along and across nomadic migration routes have long been about exchange and distribution of information on a collective and personal level at the same time. Every nomad knows that *tundra mail*[6] travels faster than people, announcing news about the arrival or departure of people, goods or events. This is to the extent that sometimes it is hard to believe that people so remote already know, for example, about my arrival beforehand. Remarkably enough, this tundra mail system has always functioned informally, long before mobile phones were introduced in the tundra. Both *tundra mail* and mobile phones, unlike unidirectional forms of communication such as TV or public radio, work along the following principles:

- Every contact is among two people, with each participant being linked to two or more people, which in turn are linked to their own rings of contact. This maintains social and economic ties, linking multiple islands over hundreds of miles.
- The communication usually serves to solve fundamental spatial problems in people's lives.
- It also serves to maintain friendly relations among the inhabitants of different areas and maintain communication with trading partners over great distance. It prepares the ground for the exchange of utilitarian items (White 2011).[7]

6 *Tundrovaia pochta*: the notion that information travels from nomadic camp to camp with different visitors and travellers acting as multipliers of the original information.

7 Interestingly, the way in which technology in Nenets nomadic society has been used to enhance flows of information, travel and consumption reminds more than incidentally

The Diversity of Knowing and Moving on the Land

To return to our point of departure, all the innovations mentioned above contribute to changing approaches to using the land, space and distance in Nenets nomadic society because they are technologies of mobility. Even rubber boots and stoves enable particular ways of being mobile that had been organized very differently before the introduction to these innovations. To illustrate this, let us imagine the very typical situation of a Nenets nomadic family in need of bread in early October. The snow is thin and still wet, and the next bakery is around 30 km away.

One way of getting the bread is by walking with the big chemical protection rubber boots. Here every step would feel the land in terms of snow texture, softness and hardness of the soil, the fragility of the river or lake ice at places too thin to carry a human foot. On the way we would see every track of animals in the snow, we may even smell the odour released by the plants when we walk on them through the thin fresh snow, and we hear the noise, e.g., of ptarmigans escaping from us. The 30 km would give an entire day of experiencing the tundra through one's own body, with the muscles tired and sucked up by the experience of the land after arriving in the village.

The other way of getting there would be on reindeer sledge. Here the little elevations, frozen hummocks in the tundra lets you feel the land even harder, but the choice of the particular route is done by the reindeer: a wise driver gives his draft reindeer freedom to choose the route to the destination (Stammler 2010: 220–21), as the thinness of the ice is felt by reindeer legs not human legs. If we are sensitive enough, we would also experience where the reindeer want to stop for a drink, or a lichen-snow-and-green-leaf snack on the way, and we would adjust our pace of travel and our smoke-breaks according to the reindeer habits. Even though human muscles get tired from holding the *khorei* (reindeer steering stick), the experience of the land will have lasted some three to four hours and have generated a different line of memory.

With the snowmobile, especially when using a newer shock-absorbed model, we would feel the slipperiness of the snow according to how much we need to pull the throttle and feel the hummocks from the crackling of the snowmobile chassis. From the relation between throttle position and our speed we can also guess our fuel consumption, which is higher in wet soft snow than in dry and solid snow. As a result, we know better how far we can travel with one fuel tank. Our smoke-breaks we would adjust according to the engine heat. If we are sensitive enough, we feel when the (usually air-cooled) engine gets too hot and needs a break for cooling down. We would not see as many tracks, would not hear the noise of animals on our way. But we would smell if the petrol in the

of Malinowski's (1920) most classical study in this field. Referring to Papua New Guinea, Bohensky et al. (2011) go as far as claiming 'the Kula Ring … is reborn in the Information Age'.

engine is burning properly, and we would hear if the engine is running smoothly, and we would feel the wind and cold more intensely due to less body movement and the higher speed and exposure. At the same time, we would wear the same chemical protection rubber boots to prevent our fur boots from getting wet from the melted snow on the warm foot rests of the snowmobile. After an hour, we would be at our destination.

The fourth way of getting the bread is by calling relatives in the village or writing a text message with an order of bread that they should bring along next time they come by snowmobile to the camp. Most likely the conversation will also include a thick description of the route between the village and the camp for the travellers from the village; it is less likely to include GPS coordinates of the camp generated through the sender's phone's GPS function or a separate GPS. In this case no travel is involved by the family who needs the bread, but still knowledge of the land to produce a narrative of the trip for others to use. This is only possible on the basis of extensive experience. Otherwise it would not make sense.

These different flows of information and goods for the same purpose show us how technological change affects our perception of the land, of space surrounding us, our idea of time and distance, and our skills and knowledge related to particular ways of movement. I would not undertake any hierarchical classification as to which sort of travel or knowledge of the land is better, more elaborate and valuable. Instead, I highlight that each way requires a particular skill set related to how we make sense of space and movement, starting from the skill of walking and observing the texture under our legs, to observing the land through reindeer behaviour, to feeling the snowmobile with its interplay between chassis and engine, to our skills of narrating travel via mobile phone. All of these skills involve conversations between humans and the land and about the land, from which anthropologists can sense and understand the ways in which ideas and practices in the environment change with the use of innovations. I would also suggest that we can do this best if we do not only participate in these practices as observers, but as agents on our own. In the cases mentioned above this would mean driving the reindeer sledge and the snowmobile yourself rather than being a passenger, and describing the route in a text message to people in the villages rather than observing how your host does so. On the basis of our own active participation it is likely that we get a deeper understanding than by passive participating through observing.

The fact that reindeer nomads did not invent the rubber boot nor the snowmobile nor the mobile phone nor the stove in the first place does not mean that they are not creative or inventive. They made the innovations diffuse among their nomadic society by seeing particular affordances in them. For this, the interplay of cognitive activity, experience and openness to introducing new items to their own culture is a precondition. At some first point in time, they must have seen the innovation used by somebody from outside their society – or at least known about the existence of these items. Afterwards they may have seen the use with their own eyes, and later tried themselves, much in the way outlined in Rogers's study on the diffusion of innovations (2003).

The crucial intellectual process involved here is seeing the affordances of these innovations for purposes which were not intended by those who invented them. However, I argue that only after having experienced the use, people start adopting these things for their own purposes. User experiences obviously vary and therefore result in different affordances being *seen*: the use of a mobile gas stove for recreational purposes is experienced differently from the use in a mobile home, a *chum*, because the requirements differ. The difference in these experiences with using technologies of mobility will also structure perceptions of space: for herders the scale of movement and the imagination of the vastness of the land may not increase with snowmobile use. On the contrary: with reindeer they used to travel further. Yet for sedentary villagers, even in the same area, the range of action and perception may indeed increase enormously. Even in the light of heavy academic arguments, it therefore seems to me that dwelling, sensual experience and cognition are inseparably combined rather than mutually exclusive, an argument supported by Ingold himself, who strongly advocates that bodily and mental experience must not be separated (Ingold *versus* Howes 2011: 325–6). This intertwining of ideas and experience in the process of adopting innovations in a society for me also highlights the problem with categorizing the diffusion of these innovations in stages. Firstly, the example with the bread mentioned above points to the fact that an innovation does not always replace a previously common way of doing things. Secondly, while Rogers's stages are useful as analytical constructs to understand some principles of how technology enters a society, we need to experience the processes of such implementation with the people to understand how it changes their perception of time and place. I would argue that it is only flows of diffusions that are experienced, while stages are constructed afterwards to understand cognitively.

Speaking about the affordances of technological innovations, we should also consider the *hardware* and *software* (see theoretical part) aspects of these, with the hardware being largely determined by the *sending* society of the innovation. In other words, the range of physically possible uses of a mobile phone or snowmobile is predetermined by the people who invented and designed them, who are in turn socialized within their respective (usually Western) setting. Each import also conveys a message from the sending culture to the receiving culture, even though neither *senders* nor *receivers* may even be aware thereof. From this follows that the consequences of introducing new technologies to a large extent are unintended and unforeseen – a conclusion that is neither new nor unique to a northern reindeer herding setting (Naimi and French 2010).

Obviously, consequences of each new innovation should be studied separately because the fields of influence are so different. This is beyond the reach of this chapter. Two last thoughts in this discussion may still be worth highlighting: I would suggest that along Ingold's line of argumentation (2005), all technologies of mobility in the North are conversations. Secondly, I would emphasize the difference between technologies of mobility involving physical movement and those involving the movement, or better say the transmission, of information. Both the snowmobile and the mobile phone can be seen as conversations or communications. The conversations here include all the features and affordances of the devices: inbuilt GPS communicates your position, inbuilt photo and video communicates events, experience to others. Everything is about travelling information and objects or subjects across space. The snowmobile originates from a Western idea of moving to destinations, rather than travelling and experiencing on the way. The mobile phone was invented in search for ever faster efficiency in information transmission, not even in flows but in packages, in unspecified locations. This is a radical difference from speaking and writing, where there is a flow of information and the message unfolds letter by letter, word by word.

In an information package, you get a whole set of information at once, you just have to *unpack it* to digest it. Such packages of information are not unknown in societies with a tradition of conversing with the spirits, where messages and symbolic representations are also transmitted in packages. Likewise many of us may have experienced a dream lasting only some seconds or minutes, which was so packed with events that it takes half an hour to tell it in words. Accounts of people who experienced death report that an entire lifetime can pass by in seconds, faster than a fast-forwarded film. Also shamans have told that in a short trance the spirits can transmit a whole lifetime.[8] After returning from trance they take a long time to unpack and relate the experienced to this reality. From that we understand the relativity of time and location. Perhaps similar to text messages, such information is transmitted as energy fields rather than flows, which is where the most radical difference to physical movement rests.

Both the snowmobile and the phone create a space of conversation that is different from one where the only way of conversing was travelling on foot or reindeer sledge, but I argue with the mobile personal communication technology, the change is more significant. The snowmobile comes instead of muscle power for moving physical objects. The mobile phone introduces a mode of travel that before was reserved for spirits and shamans. We have seen that in northern nomadic settings such transmission is not unknown, for example in shamanic soul travels in all northern indigenous cultures, including the specific Eveny *djuluchen* (Ulturgasheva 2012). However, the introduction of mobile personal communication technologies has for the first time made this kind of packet-data information transmission available for everybody through a hardware-based technology, while the previous methods were based on culturally transmitted spiritual skills more to specialists, who in some cases use *hardware* such as drums, mirrors or bells on the costumes to enable that information transfer.

8 For example an account by shaman Aleksandr Artemiev to this author, Kolyma, September 2011.

Consequently, a fundamental difference between mobile travel technologies such as rubber boots or snowmobile and mobile communication technologies such as GPS and mobile phones is the shift from the physical to the virtual in the perception of space. Earlier I reasoned that this is related to a shift from experience-based to imagined travel in space, as I have shown for the difference between travelling for getting the bread and narrating the way on the phone for the bread to be delivered. Would an increase of the cognitive elements in people's relation to each other and to their environment, and a decrease of experienced-based relations follow from that? In other words, is that like the difference between processes in the brain generated by hearing a narrative versus experiencing physical movement? Does this mean that there is no experience involved in GPS positioning, photo-sending, text messaging, etc.? I would not agree to that because narrated, text messaged or telepathic travel is also experienced.

There is a whole experience-based dimension of such use of technology. Those who have tried to type a phone number on the tundra at a temperature of minus 45 degrees Celsius may have a memory in their fingertips while reading this sentence. Yet there is more experience involved in sending a text message: I have been with herders observing how they feel when the message leaves them, by looking at the little arrow-symbol in the phone. Once that arrow is gone, you travel in your mind with your message-packet-data to the receiver, and imagine his/her reaction, await response, while multiple experiences are processed in your mind. I thought the same when I was with herders staring at the GPS display while the device was searching for satellites to receive. One by one the satellites would pop up on the display and herders commented on the connection between their location and these objects in space. From that I would argue that this sort of communication technology is also experience-based. Like with other technologies such as the stove, the rubber boot or the snowmobile, the experience of movement – physical or mental – is filtered through the device.

Thus, technology acquires the role of a mediator for the experience of conversation – with the land, with the animals or with each other – all the constituents of space. Just like the ancient Greeks needed Hermes as a messenger to communicate with their gods, now a loving boy needs text messages and smiley symbols generated by his messenger phone to get in touch with the communication partner.

Conclusion

This chapter was an attempt to show how narratives of the innovation and adoption of mobile technologies among nomads are based on both experience and memory and change people's way of perceiving their surroundings. Using examples as different as the iron stove, rubber boots, the snowmobile and personal communication technologies (PCT), I have shown how technological innovations adopted by a nomadic society, such as that of the Yamal Nenets, influence flows of people and of communication, and with this also perception and experience of

Figure 11.3 Experiencing the packet data information transmission from a mobile phone

Source: Photograph by Florian Stammler

space, most prominently through physical processes of travel, but also imagined travel. While we know that the perception of space among Nenets nomads has been very elaborate since ancient times similar to the way Anderson (2000) describes for the neighbouring Evenki, with PCT this space acquires a different experiential character, as herders imagine how information travels in packages to satellites or to distant places on the other end of the world.

I have also shown that the idea of the 'land's end', the name that Yamal has in Nenets language, was transformed by the introduction of *on* and *off* regions of mobile coverage, re-structuring to a certain extent the ideas of the remoteness and connectedness of the land by reindeer nomads. The extent to which (especially younger) members of Nenets society have adopted such a technology-based understanding shows their openness not only for new technology but also for new understandings of space. This is – on the other hand – not surprising as we know that nomadic cultures not only in the Arctic had to display exceptionally high adaptability to all kinds of changes and hazards (Bollig 2005), and livelihoods have been far less stable and more flowing than portrayed by some ethnography focusing on ecological, social and economic equilibrium analysis.

Along similar lines, I have shown how experiencing the work of technology in societies enables us to relate flows of diffusion to changes in the perception of time and place (see the example with bringing the bread). On the other hand, the stages (not flows or experience!) of innovation diffusion introduced by Rogers (2003) allow for an analytical categorization of such processes that may be good tools for comparative analysis but are not necessarily experienced by the people themselves as such. What Rogers's model does not show either is the cultural consequences of diffusion, and the simultaneity of the different stages and categories that he introduces. We can see that technologies do not necessarily have to replace each other, but complement each other, and even more – get interconnected, for example when nomads simultaneously use fur and rubber boots in the wet season, travel by reindeer sledge and snowmobile, use gas, wood stoves and open fires, and mobile phones as well as spiritual activity for transmitting package-data information and communicating over time and space.

The above shows that all these mobile technologies do not only influence the experience of physical travel but also its imagination, and consequently the flow of movement and of information in space. To me this is impressive proof that experience, memory and imagination, practice and narrative, and finally cognition and dwelling are not mutually exclusive perspectives in the human nomadic livelihood (maybe human life altogether). Fieldwork-based research has revealed here that such concepts are not only complementary but even indispensable because experience and mental representation are intertwined (Ingold 2011: 326): in our case both are part of the one process where people experience and recognize simultaneously the affordances of technology in their particular context.

Acknowledgements

Fieldwork on the narrative part of this chapter was made possible by the ORHELIA research project funded by the Finnish Academy, decision 251111. The author thanks all reindeer herders as well as Aleksandr Nil with his team from Bovanenkovo for their friendship, support and inspiring conversations throughout the years. I also thank the two reviewers of this chapter for their challenging constructive critique and encouraging comments.

References

Anderson, D.G. 2007. Mobile architecture and social life: The case of the conical skin lodge in the Putoran Plateau region, in *Les civilisations du renne d'hier et d'aujourd'hui*, edited by S. Beyries and V. Vaté. Antibes: Editions APDCA, 43–63.

Argounova-Low, T. 2012. Narrating the road. *Landscape Research*, 37(2), 191–206.

BBC 2012. Russia's New Energy Frontier (with L. Ash, F. Stammler, and A. Kovaleva). *Crossing Continents*, BBC Radio 4, 14 May, 20:30 hours [radio broadcast]. Available at: http://www.bbc.co.uk/programmes/b01h75xt [accessed: 3 October 2012].

Beach, H. and Stammler, F. 2006. Human–Animal Relations in Pastoralism [guest editors' introduction to special issue]. *Nomadic Peoples*, 10(2), 6–30.

Behnke, R., Fernandez-Gimenez, M., Turner, M., Stammler, F. 2011. Pastoral migration: Mobile systems of animal husbandry, in *Animal Migration: A Synthesis*, edited by E.J. Milner-Gulland, J.M. Fryxell and A.R.E. Sinclair. Oxford: Oxford University Press, 144–71.

Bohensky, E.L., Butler, J.R.A. and Mitchell, D. 2011. Scenarios for knowledge integration: Exploring ecotourism futures in Milne Bay, Papua New Guinea. *Journal of Marine Biology* [Online], article ID: 504651, 11 pp. (doi:10.1155/2011/504651). Available at: http://www.hindawi.com/journals/jmb/2011/504651/abs/ [accessed: 3 October 2012].

Bollig, M. 2005. *Risk Management in a Hazardous Environment: A Comparative Study of Two Pastoral Societies*. New York: Springer.

Caron, A.H. and Caronia, L. 2007. *Moving Cultures: Mobile Communication in Everyday Life*. Montreal: McGill-Queen's University Press.

Forbes, B.C., Stammler, F., Kumpula, T., Meschtyb, N., Pajunen, A. and Kaarlejärvi, E. 2009. High resilience in the Yamal-Nenets social-ecological system, West Siberian Arctic, Russia, in *Proceedings of the National Academy of Sciences of the United States of America*, 106(52), 22041–8 [Online]. Available at: http://www.pnas.org/cgi/doi/10.1073/pnas.0908286106 [accessed: 3 October 2012].

Gusev, A.V. 2010. Vremia Ust'-Poluia [The Time of Ust'-Polui], in *Istoriia Iamala: Vzgliad iz muzeinykh khranilishch*, vol. 1, edited by N.V. Fedorova. Ekaterinburg: RPP Krik Tsentr, 15–20.

Harrison, K.D. 2007. *When Languages Die: The Extinction of the World's Languages and the Erosion of Human Knowledge*. Oxford: Oxford University Press.

Ingold, T. 2005. A manifesto for the Anthropology of the North, in *Connections: Local and Global Aspects of Arctic Social Systems*, edited by A. Sudkamp. Fairbanks: International Arctic Social Science Association, 61–71.

Ingold, T. 2011. *Being Alive: Essays on Movement, Knowledge and Description*. London: Routledge.

Ingold, T. *versus* Howes, D. 2011. Debate section: Worlds of sense and sensing the world. *Social Anthropology*, 19(3), 313–31.

Istomin, K.V. and Dwyer, M.J. 2009. Finding the way. *Current Anthropology*, 50(1), 29–49.

Katz, J.E. and Aakhus, M. 2002. *Perpetual Contact: Mobile Communication, Private Talk, Public Performance*. Cambridge: Cambridge University Press.

Kemp, W. 1971. The flow of energy in a hunting society. *Scientific American*, 225, 105–15.

Khlinovskaya-Rockhill, E. 2010. *Lost to the State: Family Discontinuity, Social Orphanhood and Residential Care in the Russian Far East*. New York: Berghahn.

Klokov, K.B. 2011. Diversity of adaptive strategies of endangered herders' communities in tundra and taiga areas of Russia, in *Histories from the North: Environments, Movements, and Narratives*, edited by J.P. Ziker and F. Stammler. Rovaniemi: University of Lapland, 60–63.

Malinowski, B. 1920. Kula: The circulating exchange of valuables in the archipelagoes of eastern New Guinea. *Man*, 20, 97–105.

Massey, D. 2005. *For Space*. London: SAGE.

Naimi, L.L. and French, R.M. 2010. The unintended consequences of technological innovation: Bluetooth technology and cultural change. *The IPSI BgD Transactions on Internet Research* [Online], 6(2), 3–13. Available at: http://www.internetjournals.net/journals/tir/2010/July/Paper%2002.pdf [accessed: 3 October 2012].

Pelto, P.J. 1987 [1972]. *The Snowmobile Revolution: Technology and Social Change in the Arctic*. Prospect Heights, IL: Waveland Press.

Rogers, E.M. 2003. *Diffusion of Innovations*. 5th Edition. New York: Free Press.

Sheldrake, R., Avraamides, L. and Novák Mobifi, M. 2009. Sensing the sending of SMS messages: An automated test. *The Journal of Science and Healing* 2009(5), 272–76.

Stammler, F. 2002. Success at the edge of the land: Present and past challenges for reindeer herders of the West-Siberian Yamal-Nenets Autonomous Okrug. *Nomadic Peoples*, 6(2), 51–71.

Stammler, F. 2005. *Reindeer Nomads Meet the Market: Culture, Property and Globalisation at the End of the Land*. Münster: LIT.

Stammler, F. 2009. Mobile phone revolution in the tundra? Technological change among Russian reindeer nomads. *Folklore*, 41, 47–78 [Online]. Available at: http://www.folklore.ee/folklore/vol41/stammler.pdf [accessed: 18 November 2012].

Stammler, F. 2010. Animal diversity and its social significance among Arctic pastoralists, in *Good to Eat, Good to Live with: Nomads and Animals in Northern Eurasia and Africa*, edited by F. Stammler and H. Takakura. Sendai: Center for Northeast Asian Studies, Tohoku University, 215–43 [Online]. Available at: http://arcticcentre.ulapland.fi/docs/net_13_stammler.pdf [accessed: 3 October 2012]

Stammler, F. 2011. Oil without conflict? The anthropology of industrialisation in Northern Russia, in *Crude Domination: An Anthropology of Oil*, edited by A. Behrends, S.P. Reyna, and G. Schlee. Oxford: Berghahn, 243–69.

Ulturgasheva, O. 2012. *Narrating the Future in Siberia: Childhood, Adolescence and Autobiography among Young Eveny*. New York: Berghahn.

Vitebsky, P. 2005. *Reindeer People: Living with Animals and Spirits in Siberia*. London: Harper Collins.

Vitebsky, P. 2012. Wild Tungus and the Spirits of Places. *Ab Imperio*, 2012(2), 429–48.

White, E. 2011. *Bronislaw Malinowski: Identifying the Kula Ring of the Trobriand Islanders: The Role of Ethnographic Field Observation in Pattern Recognition*. [Santa Barbara, CA]: Centre for Spatially Integrated Social Science. [Online]. Available at: http://www.csiss.org/classics/content/98 [accessed: 3 October 2012].

Yurchak, A. 2006. *Everything Was Forever, until it Was no More: The Last Soviet Generation*. Princeton, NJ: Princeton University Press.

Ziker, J.P. 2002. Land use and economic change among the Dolgan and Nganasan, in *People and the Land: Pathways to Reform in Post-Soviet Siberia*, edited by E. Kasten. Berlin: Reimer, 207–24.

Chapter 12
From Inuit Wayfinding to the Google World: Living within an Ecology of Technologies

Claudio Aporta

When I was a child, back in Lujan de Cuyo, Mendoza, Argentina, the world unfolded in a slow number of layers, allowing me to find my own place in the world, and to gradually develop different levels of regional awareness.

A critical time in my learning was when my mother made the decision to let me walk to school on my own. It was a short walk, eight blocks in all, including crossing side streets and a large road. The walk to school over distinctively and unevenly tiled sidewalks in the early morning became my first real experience of the world, as – for the first time – I walked without being guided, held, or driven.

The walk was, in spatial terms, a line traced/threaded (Ingold 2007: 41–4) in the real world, joining two end points (the home and the school), along with many other points on the way. A sense of region/s developed in complex ways over time.

Over time, I got to know and remembered every single house on the way, the trees, the flowers, the smells, the seasonal differences. More or less gradually, the walk was expanded to other itineraries, to plazas and vineyards, to the town, to bus rides, hikes, plane rides, other cities, provinces and countries. The world unveiled itself in multiple ways, through school geography classes, television, conversations, magazines, etc. A multitude of itineraries (mine and others', present and past, real and imaginary) populated my mind and developed my geographic curiosity.

To say that my walk to school was a line is an abstraction, of course. Yet it remains true that our trajectories (the walks we walk, etc.) are essential in our construction of the world.

This chapter draws on my experience with Inuit hunters, along with observations of everyday life. As an ethnographer working on Inuit wayfinding knowledge and skills since 2000, I have shared many conversations with Inuit hunters regarding, not only their traditional methods, but also the use of maps and of the global positioning system (GPS). Over the years, I have seen how GPS units have gradually made their way from marginal navigation aids used tentatively in emergencies, to essential parts of the toolkit that most Inuit travellers take along on their trips. This chapter, however, takes the Inuit experience only as a point of departure to offer a broader reflection on how new navigational and cartographic technologies (epitomized by Google Earth and Google Maps) are actually embedded and entangled in an ever more intricate world of technologies.

It is, I argue, within this new ecology of technology that our lives (including our senses of place and region) are actually unfolding.

This chapter starts with a brief description of Inuit connections to place and space, it reflects on the changes brought by the introduction of GPS in traditional wayfinding practices, and it finally proposes the hypothesis of an ecology of technology.

How an Inuit Child Used to Develop Senses of Place and Space

Before sedentarization, Inuit children would be frequently born while travelling. Even when born in a semi-permanent or seasonal camp, the camp was in fact a point located within a broader network of trajectories (Aporta 2009). The semi-nomadic life style and the fluid composition of an Inuit community was the context in which a child's geographical and social world developed. Rasmussen (in the 1920s) noted the subtle connections between being born and travelling (Rasmussen 1976 [1929]: 47). The first time a child embarked on a trip was experienced as both a symbolic and a literal way of being introduced into the world.

An Inuit child grew up in a context of moving. Listening to stories, narratives of all kinds, in which spatial references were everywhere. Wind directions, place names, travel stories, snowdrift shapes, stars and constellations, ice formations and movement, became part of the child's experience of the world both while interacting with that world and through narratives heard in and out of the dwelling. The language itself, Inuktitut, possesses a large number of spatial connotations, to the point that the speaker, by virtue of speaking the language, must situate himself/herself (or the subject) spatially. Words and terms reflecting up/down, upstream/downstream, to/away from the shore, with/against the wind are part of the Inuktitut vocabulary (Fortescue 1988).

Children and adults lived in a world where concrete places existed within networks of precisely traced trails (Aporta 2005, 2009). In a culture where mobility was at the core of life, places made sense in the context of moving. To varying degrees, this is still the case today, as travelling remains one of the most important activities for many people engaged in harvesting practices, and in visiting neighbouring communities, almost always located hundreds of kilometres away. As the Inuit moved to permanent settlements in the 1950s and 1960s, significant changes in travel patterns occurred, but most of the Inuit continued visiting most of their traditional hunting and fishing areas.

The paths that Inuit follow (the itineraries that people undertake and remember) are not permanently registered in the landscape, but are inscribed in people's memories. Anthropologist Mark Nuttall coined the term 'memoryscape' to refer to the remembered landscape of Inuit in Northwest Greenland (1992: 39). The term refers not only to the physical territory remembered by a particular individual, but also to the community's interaction with place through time, and is rooted in earlier works of other anthropologists, such as Ingold, who states that 'it is only by virtue of his belonging to the community that a person acquires a relation to a determinate portion of natural space' (1986: 137), and Basso, who points out that through knowledge and memory hunters negotiate images and understandings of the land (1988). I would add that memoryscapes are not static entities; on the contrary, the concept expresses both the permanence of memory through time and the dynamics of people's relationship with their environments. Memoryscape is not transmitted from generation to generation as a mere corpus of geographical knowledge. As Ingold puts it, 'remembering is not so much a matter of calling up an internal image, stored in the mind, as of engaging perceptually with an environment that is itself pregnant with the past' (2000: 189).

In the case of the Inuit of the Eastern Canadian Arctic, this memoryscape is in fact extensive, complex, and precise. It extends over large expanses of territory, and is historically rooted. Even the visitor of an Inuit settlement today will observe how the Inuit resist the mobility constraints of the layouts of streets and houses, as they walk or ride their snowmobiles or all-terrain vehicles (ATV) on and off the streets, cutting through house yards, and making trails across the settlement environment, and in defiance of its design.

GIS, GPS and the Google World

After my first trip to Igloolik in 1998, it became evident that the mapping of Inuit hunters' movements and use of the land and sea would play an important role in my research. In 1999, as a graduate student in Anthropology, I took my first (and only) geographic information system (GIS) class. I soon discovered that this new realm of knowledge was completely foreign to anything I had studied, and to my experience of the world. It was extremely abstract in the extreme and counter-intuitive.

The course focused on ArcInfo, which was the standard GIS software widely taught at universities and used by industries, governments, etc. ArcInfo is part of a family of GIS programmes produced by the California-based company ESRI. GIS (beyond the ESRI family of programmes) is based on a number of principles, the most basic of which is rendering the world into three abstract geometrical forms: the point; the line; and the polygon. Another of its powerful (and abstract) principles is that the world (social, natural, cultural, linguistic, economic, political, etc.) is seen as structured and composed by data. Rendering the world into mostly quantitative but also qualitative datasets, liable to be geo referenced, is the basis of the power, versatility and ubiquity of GIS.

Those who identify GIS with mere mapping underestimate its power and do not quite appreciate its principles. GIS's strength lies in its ability to gather, store, manipulate, merge, combine, and illustrate data in spatial ways. Why should this matter to anyone whose life or profession is not concerned with GIS? The answer is that it is becoming increasingly evident that, in contemporary times, most people's lives are entangled with, and to certain degree, affected by, the increasing power of the abstract principles of GIS.

Several converging developments have made GIS very different from its early versions, and, especially, more user-friendly. These developments have also prompted GIS to change/diverge into new products and processes that have made their way into everyday life and that are changing our relationship to space, our ways to make spatial decisions, our paths in developing cognitive regions, and our knowledge of the world.[1]

A key development in this process has been the integration of GPS technology into the GIS realm. GPS connected the abstract world of GIS with the real physical world we experience. As with many other key technological developments, GPS started as a military tool (designed by the United States Department of Defense), concerned with locating targets, navigating planes and submarines, trajectory of missiles, etc.

After years of military use, it became available on the market and gradually developed into a surveying tool, a navigational aid, and finally into today's multipurpose units, through its convergence with other computerized developments, its coupling with other instruments such as Google Maps, and its use in the context of everyday driving, riding, running, sailing, walking and flying.

However user-friendly today's GPS is, its principles are also abstract. Early GPS users had to engage with the principles of Western cartography: geographic coordinates, cardinal points and, especially, lines (tracks), points (waypoints) and polygons.

Early GPS units required users to immerse into this abstraction, as they were obliged to grasp the concept of moving along lines, going to and coming from points, following arrows, working within a geographical grid, identifying cardinal points, etc. The user was, in fact, constantly reminded that he/she was actually using a geographical grid superimposed onto the world. Waypoints often had to be entered manually, and GPS units were so abstract that they had to be used in conjunction with maps or charts, and even compasses. Navigational routes were constructed upon entering key geographic coordinates, which would often be calculated from reading a map.

In the new GPS units this abstraction has disappeared, as the map is *in* the GPS. And it is not the typical static topographic map we were accustomed to. It is a zoomable map, whose perspective can also represent a horizon view, and that updates as we move. There is no need to use GPS in conjunction with an external map, and waypoints can be entered by touching the screen, without calculations or knowledge of geographic coordinates. Former navigational instruments, including astrolabes, sextants, compasses, time-keepers, sonar, charts and maps, required constant awareness of time and space, as their users had to establish relationships

1 Furthermore, GIS-based rendering of the social world is quite often presented as a kind of 'objective' and 'true' knowledge about the social world that is much more reliable than other sources of knowing.

between numbers, calculations, drawings, etc., and their location. GPS is the most revolutionary technology in the history of navigation because it offers the user the answers to any spatial question, without requiring (theoretically) any engagement with real space.[2]

GPS has become the device in which (and through which) the real world and the cartographic world join. Reality and representation are merged. A user looking at a GPS display is navigating through the representation of the world (although he/she is also moving in the real world). The merging of these technologies (GIS, GPS, computers, the internet) has resulted in a most powerful new *thing* (instrument? software? tool? How should we call it?) that we can experience in such products as Google Earth and Google Maps. The fact that the *thing* is so difficult to define shows its hybrid character. It also shows its interconnected character. The fact that its consummate form is shown in a Google product is not accident at all.

One of the most important of the Google Corporation's mission statements is 'to organise the world's information and make it universally accessible and useful'.[3] In other words, Google sees the world in terms of information liable to be organized. One of the consequences has been Google's intent to digitize virtually everything. Its founders, chief executive officers (CEO), developers, etc., see Google as the ultimate democratization tool, where everything is available to everyone. Like GIS, Google's power is to see and treat the world as data. Today, we *google* things, people, ideas, time, and places.[4] What Google created with Google Earth and Google Maps is the environment in which spatial data converge with other kinds of data. It also provided an instrument never seen before, and which is available to everyone: an almost perfect superimposition of the real world and the geographical grid.

Jorge Luis Borges created a fictional place where

> the craft of Cartography attained such Perfection that the Map of a Single Province covered the space of an entire City, and the Map of the Empire itself an entire Province. In the course of time, these extensive maps were found somehow wanting, and so the College of Cartographers evolved a Map of the Empire that was of the same Scale as the Empire and that coincided with it point for point. (Borges 1975 [1935]: 131)

Borges' short story, in which the cartographers of a fictional country created 1:1 scale maps, loses its absurdity and its paradoxical irony with Google Earth and Google Maps. The ability to zoom in to street level, and Google's quixotic goal

2 The abstraction, of course, is not 'gone' but 'hidden' in the GPS system. It is, indeed, part of the background of the technological device, which regular users have no access to.

3 See, for instance, http://investor.google.com/corporate/faq.html (accessed: 13 December 2012).

4 This phenomenon has been termed by some as 'Googlization' (Gorman and McLean 2009).

of photographing *everything* is taking geographic representation to a completely new dimension. We are living in a world where physical and social spaces have been rendered on the screen. Today, we walk about a super-documented earth with almost no secrets left.

Even the Canadian Arctic, host of so much human drama in the nineteenth century, as explorers tried to find their way to uncharted lands and sea, is being rendered to minute detail. Mining and oil companies are providing the funds to upload extremely detailed satellite images of areas where mineral exploration or exploitation is taking place. The use of satellite imagery also shows the convergence of another type of technology (space/satellite technology) into Google's powerful toolkits. *Convergence* is in fact a key concept in understanding the essence and nature of the new geographic technologies, as we live in a time where, as never before, technologies do not live or exist in isolation, but in relation to each other.

GPS and Inuit

I became fascinated with both Inuit trails and GPS at the same time, and through the same experience. During 2000, as I was doing my doctoral research following Igloolik hunters in their travels, I kept a GPS activated while moving to be able to track the places we went. Shortly after my first trips, and as I was observing the GPS data on a map on my computer, several hunters commented on the tracks, but I noticed they were talking about them as well-established trails.

A quick survey with hunters around town made it clear that most of the tracks I had been mapping were in fact traditional trails that the Inuit would break every year on fresh snow, and that followed itineraries or spatial layouts that their parents and fore parents had used themselves. It became wonderfully clear that what my GPS unit was doing was creating an inventory of people's memories. The GPS lines (constituted by a series of points) documented the ephemeral tracks left by the hunters' transportation devices on the snow. They also documented the permanent routes that people had been sharing through countless generations. By the end of the season I had amassed a large part of the Iglulingmiut's memoryscape in my computer.

As a recording device, GPS created a geometrical line (in Ingold's (2007: 43) terms, a trace) that I could plot on a map. The line on the GPS, however, was more than that. When uploaded into a computer and analysed with the proper software, the line could be deconstructed: each of the thousand points forming each track had a wealth of embedded information: latitude/longitude coordinates, speed, altitude, direction of travel, time and date.

The tremendous power of GIS consists in treating such tracks as data. In other words, GIS would not only plot a line on a map, but it will also treat each component of each point as potential parts of a dataset. The deconstruction of the trail in a series of datasets allows for a large number of other possibilities, only limited to the user's objectives and imagination.

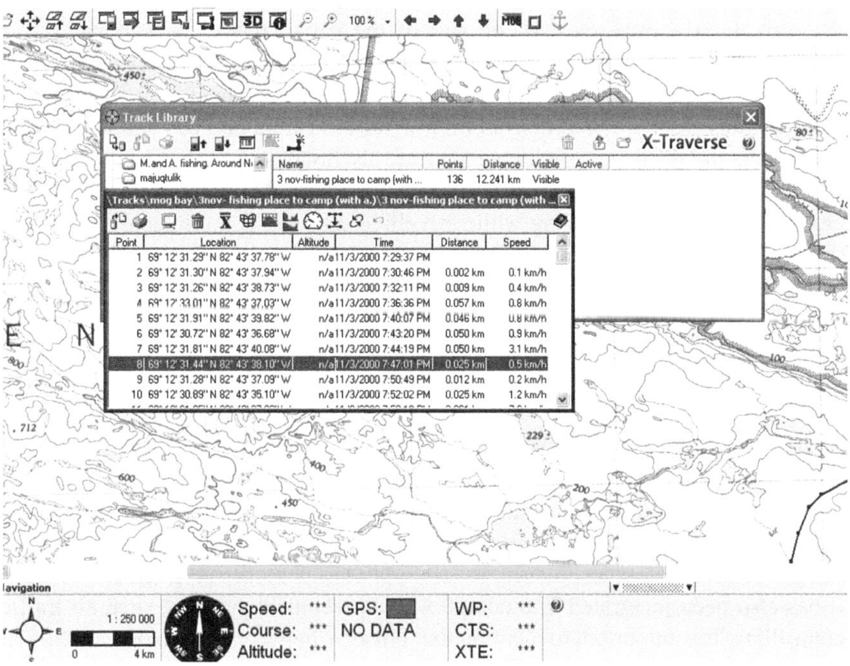

Figure 12.1 ***Inside* the data of a GPS track**

Source: Screenshot by Claudio Aporta

Grids, Networks, Convergence, and the Emergence of an Ecology of Technology

In 2005 I wrote a paper with Eric Higgs in which we concluded (along the lines with other anthropological studies of technology) that the question of whether or not GPS was changing Inuit traditional wayfinding practices was wrongfully stated (Aporta and Higgs 2005). GPS – we suggested – was to be understood in the context of use, appropriation, adoption and adaptation.

Some of the main points of our study were that:

- GPS use was not universal, there were fundamental differences between its use by older and younger hunters;
- GPS needed to be understood within the social, economic and political context of life in the settlement;
- GPS had the ability to compensate for the lower level of geographic knowledge possessed by younger generations, who (as opposed to their elders) did not have the chance of growing up on the land, and of learning traditionally.

We also pointed out, towards the end of the paper, that GPS also had to be understood in the context of other technologies, or within an *ecology of technology*. For instance, GPS has become important for hunters who travel in powerful snowmobiles, whose speed does not allow for a leisured and careful observation of the land, the wind, etc. Such machines are also loud, and their noise (along with the physical separation of driver and passengers) have made the well-established practice of knowledge sharing while travelling difficult, if not impossible. In this sense, therefore, GPS must not only be understood in connection to its social context, but also in relation to its technological context. In other words, there is a clear connection between GPS and snowmobiles. Snowmobiles are, in turn, connected to other technologies, such as gas pumps, and to the settlement itself. Pelto (1973) would have taken this further to highlight dependencies on external sources of energy.

The GPS, as a *system*, is intrinsically ecological. The GPS handheld unit is useless without its connections with satellite technologies and information technologies. It also has (through the user) technical linkages with computers and with the internet. Furthermore GPS is becoming part of other technologies, as it is integrated into cars, boats, planes, watches, phones, laptops, ski goggles, robotics. It has also been integrated into complex socio-technical systems, such air traffic controlling. It is important to mention that linkages between technologies are often established through the user, a fact that points towards a highly complex system of relations in the technological world. While the interaction between users and technologies has been noted before, the point I would like to stress here is that of the relations among technologies, through the users.

The idea proposed in this chapter is that technologies (contemporary technologies, especially) are *ecologically* related to each other. The GPS user deals with a whole structure of technologies and of social organizations. The answer to his/her concrete spatial question comes from a background of extremely complex systems where technologies interact with other technologies, and also with people, and through people.

GPS, in this sense, must be understood *ecologically*. That is, in connection with the other technologies whose ecology is shared by GPS. These technologies are actually connected to each other, and they depend on each other to exist. In the social analysis of technology, GPS, in itself, will not undermine or enrich people's perception of the world, as GPS is in fact a part of an increasingly interconnected ecosystem of technologies. In social analyses of technology, it is this ecosystem (and not individual technologies in isolation), which may affect people's lives.

A hunter's use of GPS is closely connected to the use of snowmobiles, life in town, and a new concept of time. Hunters use GPS because it affords a way to compensate for the higher speed of travel, where memorization, learning, and recognition of landmarks are more difficult or even impossible to attain. The use of GPS by Inuit hunters is also closely connected to the fact that they cannot afford to wait in bad weather (a well-established practice among pre-settlement Inuit). In foggy or whiteout conditions, hunters of the past would wait until visibility improved. Today, they need to go back to their jobs, school and town life and commitments. They also have to consider the community's anxiety generated by delays, a phenomenon that was largely absent (or that was markedly different) in pre-sedentarization times.

Hence, GPS becomes a very helpful instrument in this context, as hunters can find their way with the help of their GPS and whose ability to predict is independent of weather conditions and visibility in the real world.

I will not deny here the importance of political, economic and social contexts in analysing technologies, nor will I overlook the importance of bringing all dimensions into the analysis. However, it is my contention that there is a significant gap in analysing technology today, involving the understanding of the interconnections among technologies themselves. The relationships of technologies with each other have become so complex, so intricate, and so comprehensive, that they can be compared with the relationships established in the environmental ecological systems.

I am well aware of the shortcomings of comparisons between social and biological processes, but here I am only interested in stressing the importance, in our analyses of technology, of taking into account the networking and interconnections among devices.

Furthermore, while ecological relations among living things are taken for granted (we would not commit the absurdity of studying a plant species without its ecological context), technological connections are often absent in popular, political and academic treatments of technology. One of the main differences between the ecology of nature and the ecology of technology is that we know so little about the latter. And yet, this ecology is not metaphoric, but also physically real, and is part of our lives every day.

There have always been connections among technologies. A clear example is a composed technology of the type of a bow and an arrow. We could say that the arrow is ecologically connected to the bow, in a sense that they could both exist separately, but their functions can only be completed when both the bow and arrow (and, following Merleau-Ponty's phenomenological approach to technology (1962: 145), hand, eye and target) enter into contact.[5]

Besides these obvious technological clusters, there are other, growingly complex connections. From about the time of the industrial revolution (and particularly since the electrification of the world), connections have become more complex, to the point that establishing them becomes a task for the specialist. The history of technology shows a gradual growth of such connections, as physical grids, nets and networks have developed to encompass most of the world.

Electric or power grids, and more recently fibre optics, wireless and satellite connectivity, the internet, do not only connect people, but also connect devices among themselves (and in so doing, they make this ecology possible). This grid

5 See also Clark (2003).

allows interconnectivity, dependency, merging of devices with devices, and of devices with people. Today we live in a technological habitat.

In this picture, GIS/GPS are key developments, as they help this ecology of technology to anchor in real space (and time), and interact with the natural and social world. In the superimposed *grid* upon the real world, many answers to spatial (and many other) questions are provided – answers as general as 'east' or very specific answers, such as turning directions, speed, etc. GPS is the technology that bridges the real and the virtual. It is a technology that connects both ecologies and both worlds in our mind.

Discussion

Analysing GPS (or any other technology) in isolation offers a limited and possibly inaccurate picture of the situation. Looking at GPS in context, including its techno-ecological context, presents a more complex and serious reflection upon what is really happening. Understanding GPS ecologically will give us a better sense of the entanglement between Inuit geographic knowledge and the use of GPS/Google.

When discussing with Inuit elders and hunters in Igloolik in a workshop that took place in 2003, it became clear that the incorporation of GPS into travelling and wayfinding was both feared and welcomed. While elders warned of the danger of going through thin ice, abandoning well-established routes, and relying on an instrument that can break, freeze, or fail, they also recognized its value when used in fog, blizzards and bad visibility. They also acknowledged its potential usefulness for younger travellers.

It was recognized that GPS units were becoming part of the standard toolkit of a growing number of Inuit hunters and travellers. The elders were also enthusiastic about the possibilities of GPS, satellite imagery, etc., to map traditional land and marine use, and to help assess the sea ice in times where climatic changes hamper its predictability. Finally, they highlighted the role of this new technology as an educational tool, in classroom settings, and in connection with computers.

All the above situations in which the use of GPS was praised were connected to social and technological changes experienced in the last few decades. Implicitly, elders sanctioned the use of GPS within the new context of life, and within the new ecology of life, which technologies are an intrinsic part of.

Their reluctance and concerns were connected to two reasons: this technology does not offer a perfect representation of the world (e.g. thin ice), and technology can fail. In their assessment of GPS, the elders provided clues about the advantages and the shortcomings of this new ecology of life. Most importantly, they made this new ecology of technologies visible, through their awareness of the intrinsic connections among devices, and of the new socio-economic reality of the permanent settlement. The image of a semi-nomadic Inuit hunter, dressed in caribou clothing, riding a dog sledge, only speaking Inuktitut, living in seasonal camps, *and* using

GPS would not be conceivable. And yet, this obvious parody seems to be present when we ask whether GPS is affecting Inuit wayfinding, or how GPS is affecting our perception of the world. In the Inuit context, GPS makes sense in a broader context of other devices that have become part of the Inuit world, and in the new socio-economic and cultural reality of the permanent settlement.

Outside of the Inuit context, and in a world whose new ecology is ever growing, ever facilitating new and complex ecological relations among devices and among people, our awareness of such ecology is less clear. The *background* of technologies (to use Borgmann's term) becomes elusive, more complex and more distant, while the *foreground* (the device) is available, ubiquitous, easy and friendly (Borgmann 1984). The techno-ecological relationships established when a hunter presses a button on a GPS unit (or, for that matter, when a teenager text messages on her phone) are significantly obscure, poorly understood, and largely unrecognized.

When in 1998, an ice storm hit a large part of the Canadian provinces of Ontario and Quebec, such ecology, such network interconnecting devices and people, became visible (in its absence). The collapse of the power grid left millions of people in the dark, and provoked many dangerous situations, including more than 30 fatalities. As an unexpected side effect, the ice storm had other types of consequences. A large number of people shared their stories on national radio shows, and described how dining in the newly acquired silence of their device-free homes, with light from candles or gas and battery lamps, cooking in little gas stoves or fire places, made them rediscover connections to their surroundings and to each other that they had forgotten.

I do not wish to finish this chapter with a nostalgic appeal to pre-electricity and pre-internet days. I do, however, wish to say that many things have changed since those developments arrived in our lives, and that we do not understand them well enough.

I believe that it is fair to say that today's children may have a different experience of the world, from the one I had while walking the streets of my little town in the seventies, and as Inuit children had when learning about the winds, the snowdrifts and the stars, as they developed their perception of the world before they moved to towns.

References

Aporta, C. 2005. From map to horizons; from trail to journey: The challenges of documenting Inuit geographic knowledge. *Etudes/Inuit/Studies*, 29(1–2), 221–31.

Aporta, C. 2009. The trail as home: Inuit and their pan-arctic network of routes. *Human Ecology*, 37(2), 131–46.

Aporta, C. and Higgs, E. 2005. Satellite culture: Global positioning systems, Inuit wayfinding, and the need for a new account of technology. *Current Anthropology*, 46(5), 729–54.

Basso, K. 1988. Speaking with names: Language and landscape among the Western Apache. *Cultural Anthropology*, 3(2), 99–133.

Borges, J.L. 1975 [1935]. *A Universal History of Infamy*. London: Penguin Classics.

Borgmann, A. 1984. *Technology and the Character of Contemporary Life: A Philosophical Inquiry*. Chicago, IL: University of Chicago Press.

Clark, A. 2003. *Natural-Born Cyborgs: Minds, Technologies, and the Future of Human Intelligence*. Oxford: Oxford University Press.

Fortescue, M. 1988. Eskimo orientation systems. Meddelelser øm Grønland, *Man and Society*, vol. 11. Copenhagen: Commission for Scientific Research in Greenland.

Gorman, L. and McLean, D. 2009. *Media and Society into the 21st Century: A Historical Introduction*. Oxford: Wiley-Blackwell.

Ingold, T. 1986. *The Appropriation of Nature: Essays on Human Ecology and Social Relations*. Manchester: Manchester University Press.

Ingold, T. 2000. *The Perception of the Environment: Essays in Livelihood, Dwelling and Skill*. London: Routledge.

Ingold, T. 2007. *Lines: A Brief History*. New York: Routledge.

Merleau-Ponty, M. 1962. *Phenomenology of Perception*. New York: Humanities Press.

Nuttall, M. 1992. *Arctic Homeland: Kinship, Community and Development in Northwest Greenland*. Toronto: University of Toronto Press.

Pelto, P. 1973. *The Snowmobile Revolution: Technology and Social Change in the Arctic*. Menlo Park, CA: Cummings.

Rasmussen, K. 1976 [1929]. *Iglulik and Caribou Eskimo Texts*. New York: AMS Press.

Chapter 13

Epilogue

Tim Ingold

Are nomadic people perpetually on the move? Are sedentary people forever glued to one spot? The answer to both questions is obviously 'no'. There can be no life without movement, but nor can there be movement without rest. If the distinction between nomadic and sedentary ways of life is to retain any force, we have therefore to ask what the differences might be between nomadic and sedentary ways of moving, and between sedentary and nomadic ways of being at rest. It is curious that in all the literature on nomadic hunting and herding, while the causes and consequences of movement have been addressed at length, just what it means to move has scarcely been considered. Nor does the literature on sedentarization tell us much about what it means to rest. Under the rubric of exploring nomadic and indigenous 'spatialities', the contributors to this volume, combining perspectives from anthropology and geography, have made a start on filling this lacuna. But there is still much work to be done.

It all starts from place. For there to be places, as we learn from the many examples presented here, people have to come and go. They arrive, depart, and go around. Arriving, they converge; departing, they diverge; going around, their paths crisscross one another and they may trade greetings, news and possessions. Thus the movement of place-making, as Denis Retaillé has shown in Chapter 3, can take the forms of gathering, scattering and exchange. People may 'drop anchor' in a place, as does the mariner in a harbour, yet there can be no anchorage, Retaillé insists, without movement. But is this peculiar to nomads? Villagers have to move too, in the course of their work in farms, fields and surrounding forests. They have to bring in the hay, take livestock to pasture, cut firewood. Possibly, they have to spend more time on their feet, walking about, than do nomads. They certainly do inside the home, as Joachim Otto Habeck has shown in Chapter 8. Squatting or lolling inside the tent, the nomad expects to find all he or she needs within arm's reach, and not to have to potter about like the sedentary householder. It seems that the nomad's anchorage is, for the villager, a hive of activity. In the world of the village it is the density of this activity, and not the stillness of bodies at rest, that defines the place.

So here is another difference. Farmers have to haul themselves and their equipment against the friction of a hard and often unyielding earth, inscribing its surface with the furrows of the plough and tracks of cartwheels. Nomads, by contrast, ride the pastures much as mariners ride the waves, skimming the windswept surfaces of sand, steppe and snow, and responding in their movements, at every moment, to forces both real and imaginary (Ingold 2011: 133). Indeed it is not so much the ground surface that commands their attention as the currents of wind and weather in which they are swept up, and in which earth and sky, rather than being separated at the horizon, merge in a single field of light, sound and feeling – or in short, an atmosphere. That is how it is, too, for the animals they herd or hunt. To live, both human beings and other animals have to breathe. They breathe *in* the atmosphere, and breathe *out* along a line of movement. These lines leave their traces as tracks. Animals reveal their presence in their tracks, as do those who follow them. It is along such tracks that nomadic lives are predominantly carried on. Home for them is a path or trail, not a clearing in the forest or an oasis in the desert.

Laying and relaying their paths, nomads weave their own life-histories into the land. These trails thread their way through the land rather like the roots of a rhizome, to form a mesh of interlaced itineraries. That is why there is no contradiction between mobility and rootedness, or between nomadism and indigeneity. And it is also why sedentarization can leave people feeling rootless and bereft of connection to the land. It is like cutting a plant at the stem. There is, in this regard, a fundamental distinction between the Indigenous understanding of the land as a weave of itineraries and its state-centred or colonial representation as a territory. For in this latter conception, as Gail Fondahl has shown in Chapter 4, the land is a surface to be appropriated and apportioned rather than an ever-weaving meshwork of lines of growth and movement. This is what philosopher Gilles Deleuze and psychoanalyst Félix Guattari are getting at with their distinction between 'smooth' and 'striated' space (introduced in Chapter 1, and see Deleuze and Guattari 2004 [1987]: 408, 524–5).

The distinction is not however without its problems. In their depiction of smooth space, Deleuze and Guattari vividly evoke the atmospherics of sky and earth and the reticulate, omnidirectional pathways that are so characteristic of the nomadic lifeworld. By contrast, the peasant ploughman, having harnessed the power of horse or ox, surely striates the land with rigs and furrows, and in so doing shapes the land – creating the landscape in its original sense (as in the medieval *landskap*, see Olwig 2008). Yet the peasant is no colonist; he is not delineating a territory. If anything, he is the subject rather than the perpetrator of state-building and colonial domination. The conception of land as territory calls for a sovereign perspective; one has to see like a state (Scott 1998). Territorial lines are borders and not pathways; they define where people are by the limits within which they are confined rather than by the journeys they are on. More often than not, they cut across customary lines of travel, as Peter Koch has shown (Chapter 6) in the specific case of the historical apportionment of Lapland between nation states. Within a framework of territorial divisions and sub-divisions, such as is typically devised by the state for the efficient administration of its subjugate populations, the only way to get from place to place is 'up' (for example, from a provincial to a regional to a national centre) and then back 'down' again. The old ways, that go *along*, have been blocked.

The sovereign view of the state, according to Denis Wood in Chapter 2, is inherently cartographic. It puts the territory on the map. On the map, places are assigned to locations, defined in terms of an abstract system of coordinates, while paths reappear as the connections between them. However, the movements that actually give rise to places, and that trace the paths, are contrived to disappear. With them go the indigenous people who enact these movements and trace these paths. They, too, vanish from sight. They can nevertheless be restored to visibility through the drawing of sketch maps. In the sketch map it is the lines that count, not the points they connect or the surface on which they are drawn. To draw the lines on the map is to re-enact the movements they describe; the lines are then the traces of these movements. In theory, a sketch map could just as well be woven from threads as drawn on paper or some other surface (Ingold 2000: 241).

In Chapter 5, Nuccio Mazzullo has shown how, in a conflict between Sámi reindeer herders and the state-run Forestry Authority in Finnish Lapland, the herders were able to make their presence felt only when – with the help of the environmental organization Greenpeace – they could convert their sketch maps into maps that obeyed the strict conventions of cartography: they were to be printed, not sketched, and to include only symbols and a key, no handwriting. In Chapter 9, Joseph Long has also pointed to the incommensurability between the pastoral landscape of places and paths among Buryat people in southern Siberia and the apportioned space of state-administered territory. Here too, the discourses of conservation provided a means by which traditional Buryat experiences of landscape could be brought onto the same playing field with the media of state governance.

But new technologies are also confounding the distinction between the territorial frameworks of the state and the itinerant meshworks of once nomadic peoples. Whether among nomadic pastoralists of the Russian Arctic described by Kirill Istomin in Chapter 10 and Florian Stammler in Chapter 11, or among the Inuit hunters described by Claudio Aporta in Chapter 12, such items as snowmobiles, rubber boots, mobile phones and GPS devices have been incorporated with some enthusiasm into everyday practices of wayfaring and conversation, by people for whom experiment and improvisation have always been of the essence in coping with a fickle, inconstant and sometimes extreme environment. Every such innovation calls for the development of previously unimagined skills, while others fade away as practical opportunities for younger generations to learn them disappear. And with these new skills come different ways of perceiving the environment: freezing waters feel quite different when you are wearing boots, compared with going barefoot, and the snowmobile driver feels the snow and the ground surface in ways unlike one who is travelling by sledge. Yet technologies meant to bridge distances also create new frontiers, which are a function of the way the territorial infrastructure of the state is built into their very operation. The reach of the snowmobile, for example, is limited by the sources and availability of fuel, and that of the mobile phone by the signal range. Borders are drawn, whether on land or in the air, where none were there before.

Besides combining the disciplinary perspectives of anthropology and geography, this volume has also sought to bring together two approaches to space: one focusing on its production, the other on its cognition. Thus in Chapter 8, Joachim Otto Habeck has argued that the restructuring of spatial practices brought about by policies of settlement in the Soviet era, as they impacted on northern nomadic peoples, not only limited their movement but also led to a form of 'cognitive enclosure'. And Kirill Istomin, in Chapter 10, has insisted that among Nenets tundra dwellers of Western Siberia, finding the way from place to place involves the determination of direction, through cognitive triangulation, prior to following it through on the ground. The Nenets, according to Istomin, are consummate mental geometers who have no difficulty in conceiving of places as points in a Euclidean space, and of routes as linear connections. Routes, then, are not memorized but calculated afresh on each occasion. Among the Inuit described by Aporta in Chapter 12, quite to the contrary, paths are inscribed in people's memories, and places are known not as points in space but as loci of convergence within a pattern of interlaced trails. The contrast is stark, but does it reflect real differences of environment and culture between tundra dwellers of the Canadian and the Siberian Arctic, or rather the theoretical stances of their respective ethnographers?

There is a tension, here, between cognitivist and phenomenological approaches that is never fully resolved. Of course, there must be a place in every account both for thinking and doing, for intellectual speculation and for practical, environmental engagement. People everywhere travel in their imaginations as well as on the ground. The real question is one of ontological priority. Is thinking the activity of a mind that already dwells in the world, or are we to regard what people do, their practical activities, as the outward behavioural expression of cognitive processes centred on the brain? We cannot have it both ways. Brian Donahoe, in Chapter 7, has argued strongly for the latter view; my own preference is for the former (Ingold 2000: 219–42). No one would deny the importance and potential relevance of understanding how brains work. Brains, however, do not go hunting, nor can you find them out on the Arctic tundra or in the African savannah looking after herds of animals. These are things that people do. Crucially, they involve knowing in movement, of an animate body that is alive to its environment. There is a risk, clearly articulated by the philosopher Maxine Sheets-Johnstone, that 'brain-tethered explanatory statements about human experience overlook completely the essentiality of movement to our knowledge of the world' (Sheets-Johnstone 2011: 119). Indeed, neuro-imaging or brain-mapping does to the mind precisely what cartography does to the world: it contrives to make movement disappear. And if movement disappears, so does its corollary: stillness or rest. The map is static, but the equilibrium of rest, like a musical pause, is dynamic.

The production of space, then, cannot be separated from the cognition of space, nor do these afford complementary perspectives. Our concern must rather be to understand the dynamics of thinking-in-movement and, by the same token, of thinking-at-rest. It is from these dynamics that human beings and other animals, whether nomadic or sedentary, find their place in the world.

References

Deleuze, G. and Guattari, F. 2004 [1987]. *A Thousand Plateaus: Capitalism and Schizophrenia*, translated by B. Massumi. London: Continuum.

Ingold, T. 2000. *The Perception of the Environment: Essays on Livelihood, Dwelling and Skill*. London: Routledge.

Ingold, T. 2011. *Being Alive: Essays on Movement, Knowledge and Description*. London: Routledge.

Olwig, K. 2008. Performing on landscape versus doing landscape: Perambulatory practice, sight and the sense of belonging, in *Ways of Walking: Ethnography and Practice on Foot*, edited by T. Ingold and J. Lee Vergunst. Aldershot: Ashgate, 81–91.

Scott, J.C. 1998. *Seeing Like a State: How Certain Schemes to Improve the Human Condition Have Failed*. New Haven, CT: Yale University Press.

Sheets-Johnstone, M. 2011. The imaginative consciousness of movement: Linear quality, kinaesthesia, language and life, in *Redrawing Anthropology: Materials, Movements, Lines*, edited by T. Ingold. Farnham: Ashgate, 115–28.

Index